T0365955

Offbeat Physics

Offbeat Physics

Machines, Meditations and Misconceptions

Edited by
P. I. C. Teixeira

CRC Press is an imprint of the
Taylor & Francis Group, an **informa** business

First edition published 2022
by CRC Press
6000 Broken Sound Parkway NW, Suite 300, Boca Raton, FL 33487-2742

and by CRC Press
4 Park Square, Milton Park, Abingdon, Oxon, OX14 4RN

CRC Press is an imprint of Taylor & Francis Group, LLC

Library of Congress Cataloging-in-Publication Data

Names: Teixeira, Paulo Ivo Cortez, 1965- editor.
Title: Offbeat physics : machines, meditations and misconceptions / edited by P.I.C. Teixeira.
Description: First edition. | Boca Raton : CRC Press, 2022. | Includes bibliographical references and index.
Identifiers: LCCN 2021049576 | ISBN 9781032021164 (hardback) | ISBN 9781032033938 (paperback) | ISBN 9781003187103 (ebook)
Subjects: LCSH: Physics. | Machinery, Dynamics of.
Classification: LCC QC23.2 .O44 2022 | DDC 530--dc23/eng/20211201
LC record available at https://lccn.loc.gov/2021049576

ISBN: 978-1-032-02116-4 (hbk)
ISBN: 978-1-032-03393-8 (pbk)
ISBN: 978-1-003-18710-3 (ebk)

DOI: 10.1201/9781003187103

Publisher's note: This book has been prepared from camera-ready copy provided by the author.

Typeset in CMR10
by KnowledgeWorks Global Ltd.

This book is dedicated to Maria Helena Cardoso, to whom the Physics Department at ISEL owes its inception.

Contents

Preface

This book is a collection of articles written by past and present academic staff members of the Physics Department at the Lisbon Engineering School (ISEL) of the Polytechnic Institute of Lisbon (IPL). Many of these articles, but not all, have been published in refereed international physics education journals such as the American Journal of Physics, the European Journal of Physics, The Physics Teacher, etc. A few have appeared in Gazeta de Física, the on-membership magazine of the Portuguese Physical Society. Fewer still are original, unpublished results.

Chapters/articles cover a wide variety of topics in classical physics. This is because they are inspired by our teaching: we teach engineering students basic mechanics, electromagnetism and thermodynamics. Ultimately, all chapters originated from issues we came across while researching or writing up courses, or were prompted by our students' questions, for which we could find no satisfactory answers in the literature. In other words, the problems addressed may be basic, but they are far from trivial.

The book is divided into three parts. Part I—Machines comprises chapters that explain or model the workings of a number of machines (understood in a broad sense) on the basis of physical principles. These machines can be as simple as a rolling wheel, or as complex as a jet engine. Then in Part II—Meditations, authors attempt to go beyond the standard examples, experiments and approximations discussed *ad nauseam* in most physics textbooks, but which are not always very exciting or realistic. For example, what happens when colliding bodies are not perfectly rigid—as we know real bodies are not? Finally, Part III—Misconceptions aims to correct misconceptions that students may have about physical phenomena, or clarify issues that are often presented misleadingly, confusingly or imprecisely in textbooks, such as the relationship between angular momentum and angular velocity in rotational motion.

Our target readership is anyone with an interest in physics who wishes to learn more about this fascinating science. In particular, teachers and instructors at all levels, who often have to field unexpected and disconcerting questions, as well as highly motivated undergraduate students taking General Physics courses, may find this book useful. We hope that the realistic problems we tackle, drawn as they are from engineering contexts, will help them teach and learn physics, both theretically and experimentally.

The Physics Department at ISEL was founded by Professor Maria Helena Cardoso, who hired most of the authors of this book, under difficult circumstances. She advocated excellence in research and teaching, including good laboratory facilities. As such, she was the facilitator of the work presented here, and we dedicate this book to her.

P. I. C. Teixeira (Editor),
ISEL—Instituto Superior de Engenharia de Lisboa,
Instituto Politécnico de Lisboa,
Lisbon, Portugal, July 2021.

Acknowledgments

We are grateful to the ISEL—Lisbon Engineering School for funding, and to the American Association of Physics Teachers, the American Institute of Physics, the Institute of Physics Publishing, John Wiley and Sons, and Sociedade Portuguesa de Física, for kind permission to re-use materials originally published in their journals.

Contributors

A. C. Barroso
Departamento de Matemática, Faculdade de Ciências da Universidade de Lisboa
Lisbon, Portugal

A. Casaca
ISEL – Instituto Superior de Engenharia de Lisboa, Instituto Politécnico de Lisboa
Lisbon, Portugal

A. M. Nunes
Departamento de Física, Faculdade de Ciências da Universidade de Lisboa
Lisbon, Portugal

V. Oliveira
ISEL – Instituto Superior de Engenharia de Lisboa, Instituto Politécnico de Lisboa
Lisbon, Portugal

P. Patrício
ISEL – Instituto Superior de Engenharia de Lisboa, Instituto Politécnico de Lisboa
Lisbon, Portugal

J. P. Silva
Departamento de Física, Instituto Superior Técnico, Universidade de Lisboa
Lisbon, Portugal

A. J. Silvestre
ISEL – Instituto Superior de Engenharia de Lisboa, Instituto Politécnico de Lisboa
Lisbon, Portugal

J. M. Tavares
ISEL – Instituto Superior de Engenharia de Lisboa, Instituto Politécnico de Lisboa
Lisbon, Portugal

Part I

Machines

1

Dynamics of Braking Vehicles: From Coulomb Friction to Anti-Lock Braking Systems

J. M. Tavares

CONTENTS

The dynamics of braking of wheeled vehicles is studied using the Coulomb approximation for the friction between road and wheels. The dependence of the stopping distance on the mass of the vehicle, on the number of its wheels and on the intensity of the braking torque is established. It is shown that there are two regimes of braking, with and without sliding. The advantage of using an anti-lock braking system (ABS) is put in evidence, and a quantitative estimate of its efficiency is proposed and discussed.

1.1 Introduction

In introductory textbooks of Physics, friction is taught in order to allow applications of Newton's second law to simple mechanical systems (e.g. [1]). The

Reproduced from J. M. Tavares, "Dynamics of braking vehicles: from Coulomb friction to anti-lock braking systems," European Journal of Physics **30**, 697–704 (2009), https://doi.org/10.1088/0143-0807/30/4/004, with the permission of IOP Publishing Ltd.

DOI: 10.1201/9781003187103-1

ubiquitous example in such books is the one of blocks, pushed, pulled or moving freely in horizontal or inclined rough surfaces. However, students, when exposed to these concepts, immediately seek other examples, closer to their own experience and interest. One of these, especially if they want to become mechanical engineers, is the braking of vehicles.

The application of the simple Coulomb relation for kinetic friction to braking vehicles may explain some of its features (like the dependence on road conditions), but it leaves many questions to answer (like mass and braking torque dependence). Moreover, this relation applies when the vehicle is sliding (i.e. when the wheels are locked), and most students are aware that this condition is never fulfilled in modern cars, equipped with anti-lock braking systems (ABS).

The explanation of mechanical engineers for braking relies on a different model for friction [2]. A velocity-dependent phenomenological friction coefficient is adopted [2, 3], and all the main features of braking are obtained. However, this model can only be solved numerically (by solving an highly non-linear system of differential equations): relations between the parameters of the model (friction coefficients, mass, braking torque, inertia of the wheels) and its outcome (stopping distance, accelerations, etc) evolve the test of a considerable number of cases for the parameters.

In this article I show that these difficulties may be overcome and a more clear picture of braking can be provided in introductory courses of mechanics, if one combines the dynamics of rolling with the usual Coulomb model for friction (kinetic and static). This simpler approach has its origin in the discussion of [3] with my students. In fact, that article uses the usual approximation of engineers to road-tire friction (through a velocity dependent friction coefficient) and introduces the ABS through a new term in the equation of motion that is proportional to the derivative of the angular acceleration of the wheel. These are much better approximations to reality than the ones proposed here, but I found that such approximations diverted the attention of students (first-year Mechanical Engineering undergraduates) from the physics to the mathematical details. After some thought, I realized that, at the cost of using some less realistic hypothesis, the mathematics can be simplified while the physics still retains the main features of braking.

This paper is organized as follows. In Sec. 1.2, Newton's equations for braking are derived, and considering friction given by Coulomb's model, the dependence of the stopping distance on several parameters is established. It is shown that there are two regimes for braking and that the stopping distance can be reduced if some control over the braking torque is introduced. In Sec. 1.3, an approximate quantitative description of such a control system—the ABS—is provided. This allows the prediction of the conditions under which the stopping distance is decreased by the use of ABS. Finally, in Sec. 1.4, a comparison with the model of [3] and with real data is done, and some conclusions are drawn.

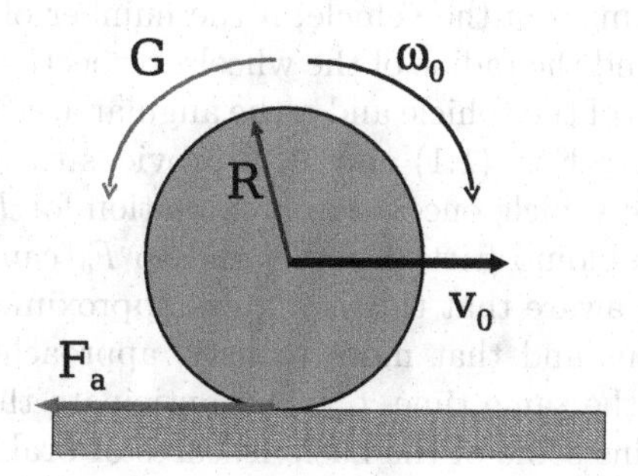

FIGURE 1.1
Schematic representation of a wheel of a vehicle that is moving horizontally and braking. G is the braking torque, F_a is the friction force, v_0 is the center of mass velocity of the vehicle, and ω_0 is the angular velocity of the wheel. For simplicity, we do not represent in this figure the weight of the vehicle (per wheel) and the road normal reaction.

1.2 The dynamics of braking using Coulomb friction

Consider a vehicle moving in a straight horizontal line with center of mass velocity v_0. The wheels of the vehicle are spinning with an angular velocity ω_0. When the driver pushes the braking pedal, a braking torque G is applied on each wheel and a friction force of magnitude F_a between the wheel and the road appears. It is this force that will make the vehicle stop. In Fig.1.1 we represent schematically the initial instant of action of braking in a wheel. To simplify our approach we consider that the road is perfectly horizontal, the weight is equally distributed through all n wheels, the wheels are rigid and acted by a friction force of the same magnitude, and that G is constant and includes the effect of all torques (break, engaged clutch, etc.) that decelerate the rotation of the wheel. Within these approximations, the equations of motion for the translation of the vehicle and for the rotation of the wheels are,

$$Ma = -nF_a, \tag{1.1}$$

and

$$I\alpha = -G + RF_a, \tag{1.2}$$

where M is the mass of the vehicle, n the number of wheels, I and R the moment of inertia and the radius of the wheels, respectively, a the acceleration of the center of mass of the vehicle and α the angular acceleration of the wheels.

The integration of Eqs. (1.1) and (1.2) provides the complete description of the motion of the vehicle once an approximation for F_a is adopted. In this work we use the Coulomb friction force, and so F_a can be static or kinetic. Students should be aware that this is a gross approximation for the complex road-tire interactions and that more realistic approaches are available (like those of [2, 3]). At the same time, one can anticipate that, nevertheless, this simpler model retains some of the basic features of braking.

1.2.1 Static friction force

If at all instants the velocity of the car v and the angular velocity of the wheel ω are related by $v = R\omega$ (i.e. $a = R\alpha$), then the wheel is rolling perfectly (i.e. without slipping or skidding) and the friction force is static. The combination of Eqs. (1.1) and (1.2) with the condition for rolling $a = R\alpha$, gives the acceleration of the car, the angular acceleration of the wheels and the value of the friction force,

$$a = -g\frac{\Gamma}{1+\nu}, \tag{1.3}$$

$$\alpha = -\frac{g}{R}\frac{\Gamma}{1+\nu}, \tag{1.4}$$

$$F_a = \frac{G}{R}\frac{\nu}{1+\nu}, \tag{1.5}$$

where g is the acceleration of gravity, Γ is a reduced torque, $\Gamma \equiv \frac{GR}{Ig}$, and ν a reduced inertia [1], $\nu \equiv \frac{MR^2}{nI}$. Since the friction force is static then $F_a \leq \frac{\mu_s Mg}{n}$, μ_s being the static friction coefficient between the wheel and the road. Therefore, using Eq. (1.5), braking with static friction can only happen if,

$$\frac{\Gamma}{1+\nu} \leq \mu_s. \tag{1.6}$$

If the applied braking torque G is constant, then the acceleration a is also constant, and the stopping distance is $d = \frac{v_0^2}{2a}$, where v_0 is the initial velocity. Notice that we are neglecting the so-called reaction distance, i.e. the distance traveled by the vehicle between the instant when the driver realizes he has to brake and the instant when G is established. Using Eq. (1.3), one obtains for d_s, the stopping distance in the case of static friction,

$$d_s = \frac{v_0^2}{2g}\frac{1+\nu}{\Gamma}. \tag{1.7}$$

[1]Notice that ν is approximately the ratio between the mass of the vehicle and the mass of all wheels. For the most common vehicles (like cars and trucks), $\nu \gg 1$, and so, in all expressions one could use $\nu + 1 \approx \nu$.

This equation shows immediately that an increase in the braking torque, a decrease in the mass of the vehicle or an increase in the number of wheels all decrease the stopping distance. Moreover, it shows that, until a certain limit defined by Eq. (1.6) drivers can fully control the braking of their vehicle by just pushing more or less the braking pedal.

1.2.2 Kinetic friction force

When the wheels are not rolling (i.e. $v \neq R\omega$) or condition (1.6) is not obeyed, the friction force is kinetic,

$$F_a = \mu_k \frac{Mg}{n}, \tag{1.8}$$

where μ_k is the kinetic coefficient of friction between the wheel and the road. The acceleration of the vehicle and the angular acceleration of the wheels are,

$$a = -\mu_k g, \tag{1.9}$$

$$\alpha = -\frac{g}{R}(\Gamma - \mu_k \nu). \tag{1.10}$$

Notice that in this case the acceleration a is independent of the applied torque G, so the driver has no control over braking. The stopping distance is,

$$d_k = \frac{v_0^2}{2g\mu_k}. \tag{1.11}$$

Moreover, when the friction force is kinetic, the lateral static friction force decreases strongly. This decrease makes spinning of the vehicle easier, specially if the trajectory is not linear. The present model (like the one in [3]), given the strong assumptions done for the direction of the motion and of the forces, does not account for this effect. Nevertheless, prevention of spinning should be referred to students as the main advantage of having a anti-lock system.

1.2.3 The two regimes for braking

Consider the following situation: a vehicle is rolling ($v = R\omega$) in a horizontal road with velocity v_0 and, at instant $t = 0$, a constant braking torque G is applied at each wheel. Further assume that, as usual, the static friction coefficient is larger than the kinetic one ($\mu_s > \mu_k$).

Let us investigate the motion of the vehicle and of its wheels for different values of G. If G satisfies Eq. (1.6) then the acceleration depends on G and can be controlled by the driver; the vehicle will keep rolling ($v = R\omega$ and $a = R\alpha$) until it stops. On the other hand, if G is large enough and does not satisfy Eq. (1.6) then the acceleration does not depend on G. Equations (1.9) and (1.10) can be used to conclude that $R|\alpha| > |a|$, so that the angular velocity of the wheels decreases faster than the velocity of the vehicle: the wheels will lock

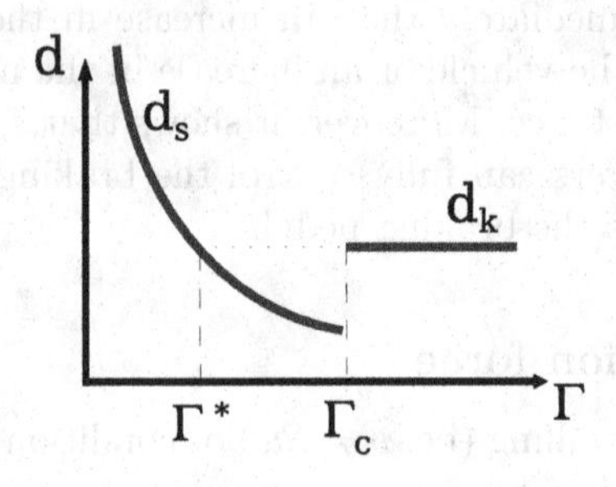

FIGURE 1.2
Stopping distance d as a function of reduced braking torque Γ, when a constant braking torque is applied to a rolling ($v = R\omega$) vehicle. The stopping distances in the static friction regime, d_s, and in the kinetic friction regime, d_k, are given by Eqs. (1.7) and (1.11), respectively. Γ_c is Eq. (1.12) and $\Gamma^* = \mu_k(1+\nu)$. In the static friction regime, the stopping distance can be controlled by varying the braking torque. For values of Γ in the range $\Gamma^* < \Gamma < \Gamma_c$ one has $d_s < d_k$.

before the vehicle stops. Therefore, Eq. (1.6) can be used to define a critical value for the reduced braking torque,

$$\Gamma_c = \mu_s(1+\nu), \tag{1.12}$$

that defines two regimes for braking, given an initial condition of rolling: (i) $\Gamma \leq \Gamma_c$ braking with a static friction force; (ii) $\Gamma > \Gamma_c$: braking with a kinetic friction force. Notice that Eq. (1.12) is a limiting case of a more sophisticated model (Eq.(10) of [3]).

In Fig.1.2 we plot schematically the dependence of the stopping distance on Γ for these two regimes. One can easily see that it is possible to have control over the braking (no lock of the wheels) and a shorter stopping distance than in the kinetic friction case, if the applied reduced braking torque Γ satisfies $\mu_k(1+\nu) < \Gamma < \Gamma_c$. As a consequence, to brake better (i.e. in less distance and with more control over the vehicle) is not to brake more, but to choose the adequate level of braking. The next section explains how this choice can be accomplished.

1.3 The advantage of the ABS

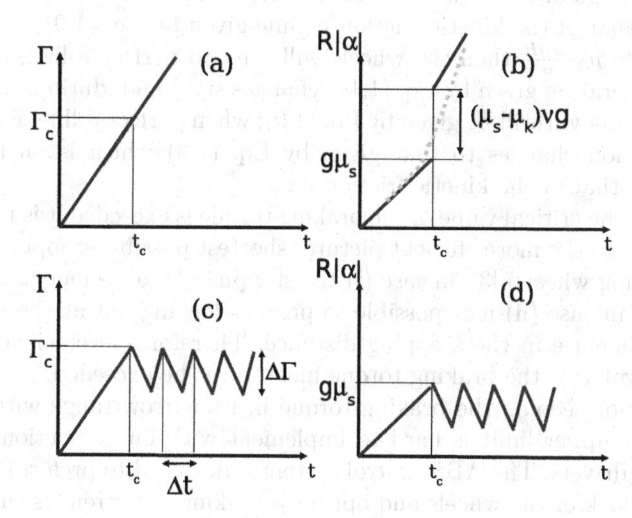

FIGURE 1.3
Applying a linearly increasing braking torque without ((a) and (b)) and with ((c) and (d)) ABS. (a) The reduced braking torque Γ increases with time t and reaches the critical value Γ_c given by Eq. (1.12) at $t = t_c$. (b) According to the model under study, the angular acceleration α increases also linearly, but, at $t = t_c$, there is a large discontinuity in α—see Eqs. (1.4) and (1.10)—that corresponds to the change in the braking regime; the dotted line represents the strong variation of α in real situations, that is detected by the ABS. (c) We consider that the strong variation of α seen in (b) is detected by the ABS at $t = t_c$; after this instant, the ABS takes control over the braking by, in cycles of duration Δt, ordering variations of $\pm\Delta\Gamma$ in the braking torque. (d) The strong variations of α are detected by the ABS and give origin to the orders of increasing/decreasing the braking torque.

The need of a control system to obtain a more efficient braking can be put in evidence by considering a driver trying to brake the vehicle (not equipped with ABS) in the most efficient way (or at least more efficiently than just pushing hard the pedal and let the wheels lock). The driver must choose a pressure in the pedal to create a reduced torque Γ in the range $\mu_k(1+\nu) < \Gamma \leq \Gamma_c$, or, using the definition of Γ and the approximation $\nu+1 \approx \nu$, a torque G in the range $\mu_k \frac{MgR}{n} < G \leq \mu_s \frac{MgR}{n}$. This is usually a narrow range that depends on the friction coefficients. Moreover, if the driver happens to push a bit harder the pedal so that G exceeds the critical value, the wheels will start slipping and the friction force becomes kinetic. If he then reduces the braking to a value of G below the critical one, two situations may happen:

(i) if $G > \mu_k \frac{MgR}{n}$ the wheels will continue its tendency to lock (although with a lower angular acceleration—see Eq. (1.10)) and the vehicle acceleration will remain that of the kinetic friction regime given by Eq. (1.9);

(ii) if $G < \mu_k \frac{MgR}{n}$ then the wheels will tend to perfect rolling (since the angular acceleration given by Eq. (1.10) changes sign) and, during this period, the acceleration will still be given by Eq. (1.9); when perfect rolling is attained, the acceleration changes to that given by Eq. (1.3) which is, in this case, smaller than that of the kinetic friction regime.

So, when the critical value of the braking torque is exceeded it is impossible to come back to the more efficient picture (shortest possible stopping distance and nonlocking wheels) [3]: in case (i) the stopping distance increases and the wheels lock; in case (ii) it is possible to prevent locking but at the cost of an even large increase in the stopping distance. Therefore, in efficient braking, the critical value of the braking torque must never be exceeded.

The task of keeping the braking torque in its narrow range without ever exceeding its upper limit is hard to implement with human action, at least for common drivers. The ABS control system is designed to preform this task: prevent the lock of the wheels and optimize braking in extreme situations.

The way this control system works can be understood in the context of the model presented in the previous section. Suppose that a vehicle is rolling and that, at some instant, an increasing braking torque is applied (see Fig.1.3). When the critical braking torque is reached, the angular acceleration of the wheels is $R|\alpha| = g\mu_s$ (obtained from Eqs. (1.4) and (1.12)); but, in the next instant, the transition to the kinetic braking regime happens, and $R|\alpha| = g\mu_s + g(\mu_s - \mu_k)\nu$ (obtained from Eqs. (1.10) and (1.12)). Therefore, a strong discontinuity in the angular acceleration of the wheels is predicted by this model when the change between the two braking regimes takes place. In real systems, this discontinuity corresponds to a strong increase in $|\alpha|$. The ABS detects this strong increase and orders the braking torque to decrease, anticipating the change of braking regime. But when the braking torque decreases rapidly, $|\alpha|$ decreases also rapidly. The ABS detects this strong decrease and orders the braking torque to increase.

In engineering models the ABS is included in Eq. (1.2) by considering G proportional to the time derivative of α [2, 3]. Here we propose a simpler approximation within the Coulomb friction model. We consider that the ABS orders the decrease in braking torque precisely when $\Gamma = \Gamma_c$. After a decrease of $\Delta\Gamma$ [2] the braking torque increases again until $\Gamma = \Gamma_c$ is reached. This cycle is assumed to have a time duration Δt, and to repeat until the vehicle stops (see Fig.1.3). Moreover, Δt is assumed to be small when compared with the

[2]One must have $\Delta\Gamma \leq \Gamma_c$, *i.e.* the braking torque that results from the order of the ABS to decrease the torque can be 0 but cannot change sign (otherwise it would not be anymore a braking torque).

time to brake [3] [4], so that the acceleration of the vehicle when the ABS is working may be approximated by a mean acceleration, a_{ABS},

$$a_{\mathrm{ABS}} = -\frac{g}{1+\nu}\left(\Gamma_c - \frac{\Delta\Gamma}{2}\right). \tag{1.13}$$

This expression shows that the ABS system gives a sub-optimal deceleration, since $|a_{\mathrm{ABS}}|/g < \frac{\Gamma_c}{1+\nu}$, and explains the reason why a very experienced (usually racing) driver can brake better even without ABS.

From the constant acceleration given by Eq. (1.13) the stopping distance with the use of ABS, d_{ABS}, can easily be computed and compared with that of the kinetic friction braking regime defined by Eq. (1.9), to obtain,

$$\frac{d_k}{d_{\mathrm{ABS}}} = \frac{\mu_s}{\mu_k}\left(1 - \frac{\Delta\Gamma}{2\Gamma_c}\right). \tag{1.14}$$

This ratio can be used as a measure of the efficiency of the ABS. Using Eq. (1.12), Eq. (1.14) becomes,

$$\frac{d_k}{d_{\mathrm{ABS}}} = 1 + \frac{\mu_s - \mu_k}{\mu_k} - \frac{1}{\mu_k}\frac{\Delta\Gamma}{2(1+\nu)}. \tag{1.15}$$

This equation leads to the schematic representation of Fig.1.4. In fact, we have separated the parameters that depend on the properties of the vehicle ($\frac{\Delta\Gamma}{2(1+\nu)}$) from those that depend on road conditions (μ_k, μ_s). This separation allows a clear prediction of the conditions under which the ABS will be efficient. In Fig.1.4, we represent schematically Eq. (1.15): when $\mu_s > 2\mu_k$ one has $d_{\mathrm{ABS}} < d_k$, independently of the characteristics of the vehicle (i.e. of $\frac{\Delta\Gamma}{1+\nu}$); when $\mu_s < 2\mu_k$ the ABS will only be efficient if $\frac{\Delta\Gamma}{2(1+\nu)} < \mu_s - \mu_k$. Therefore, we can conclude that for the ABS to decrease the stopping distance under most road conditions, it has to be designed so that $\Delta\Gamma$ is small. In fact, in some commercial ABS systems [4], in one cycle, the release of brake lasts $\approx$ 20ms, while the consequent increase lasts $\approx$ 200ms and is formed by a sequence of plateaus, closer and closer to the limit Γ_c. This is a way to reduce the mean difference between the critical braking torque and the braking torque during an ABS cycle, or, in the language of our model, to decrease $\Delta\Gamma$.

1.4 Comparison with the model [3] and with real data

The model presented in this article can be compared with a more sophisticated one [3], where a vehicle characterized by $\nu = 15$ and a braking with $\Gamma = 20$ was

[3]The typical duration of a cycle in ABS is $\Delta t \approx 0.1$s; the typical time to brake is of a few seconds (depending on the initial velocity).

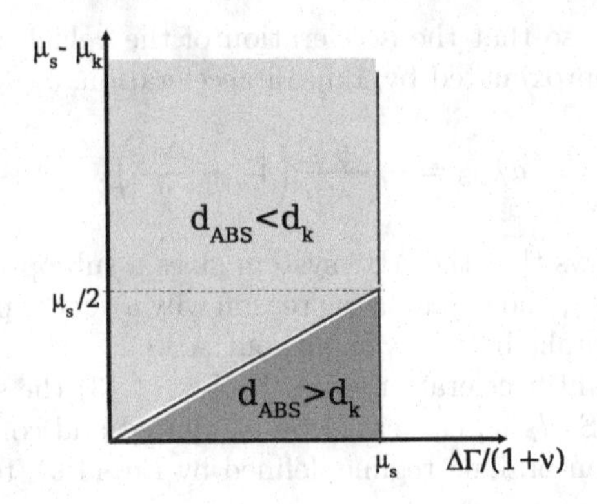

FIGURE 1.4
Ranges of the parameters of the model for which $d_{\mathrm{ABS}} < d_k$ and $d_{\mathrm{ABS}} > d_k$, using Eq. (1.15). The horizontal axis represents the dependence on the properties of the vehicle through $\frac{\Delta\Gamma}{1+\nu}$ and the vertical axis the road/wheel interaction through $\mu_s - \mu_k$. Therefore, this diagram may be applied very easily to real situations: a vehicle is a vertical line and a pair road/wheel an horizontal one.

considered. The friction force was obtained from a friction coefficient μ that depends on the ratio $\frac{v}{R\omega}$: μ has a maximum value of 0.5, and is 0.3 when the wheels are locked. During the working of the ABS, the applied braking torque can be reduced to a minimum value of $\Gamma_{\mathrm{min}} = 5$. These parameters intend to model the braking in a wet road. They can be introduced in our model by considering $\mu_s = 0.5$, $\mu_k = 0.3$, $\Gamma_c = 0.5(1+\nu) = 8$ and $\Delta\Gamma = \Gamma_c - \Gamma_{\mathrm{min}} = 3$. For an initial velocity $v_0 = 20\mathrm{ms}^{-1}$, using Eqs. (1.11) and (1.14) we compute $d_k = 67$m and $d_{\mathrm{ABS}} = 50$m, which coincide with the results of the more complex calculations of [3].

Reliable real data for stopping distances, braking decelerations and, specially, friction coefficients is hard to find. Nevertheless, we attempt here a comparison between our approach and some scattered data found over the Internet [5,6].

In [5] a list of decelerations during braking for different cars and two initial velocities can be found. The decelerations a/g range from 0.98 to 0.55, and are almost the same for each car (i.e. its dependence on the initial velocity is negligible). Moreover, the larger decelerations are obtained for "better" cars (like Ferrari, Mercedes, etc.). These data renders support to Eq. (1.3): there is no dependence on the initial velocity, and the deceleration can be increased by the car design (increasing Γ and decreasing ν).

The web page of the road authority of Queensland, Australia [6], presents a list of interesting data that compares the stopping distances in dry and wet surfaces for eight initial velocities (ranging from 50 km/h to 110 km/h). For each road conditions the v_0^2 dependence is assumed. The values of the deceleration a/g are 0.71 for dry and 0.49 for wet conditions. In wet conditions the deceleration is much smaller, and thus the sliding regime (see Eq. (1.12)) is much more likely to occur. Therefore, under these conditions the ABS system is specially convenient.

In conclusion, the model for braking developed here is simple but captures some nontrivial phenomenology of the braking of vehicles. It can be used to demonstrate to students the connection between models and the reality they experience. Students that want to graduate in Mechanical Engineering are specially interested in this problem. This motivation leads them to a deeper understanding of the dynamics and kinematics of rotation, through the study of braking. The model presented here can be the origin of student's projects (theoretical and numerical). There are several interesting generalizations that can be proposed to students, e.g.: the distinction between rear and front wheels, which support, especially during braking, different loads; the application of the same model to the acceleration of vehicles; the modeling of the friction force like in [2, 3] and the consequent numerical solving of the equations of motion.

References

[1] P. M. Fishbane, S. Gasiorowicz, and S. T. Thornton, *Physics for Scientists and Engineers*, 2nd ed. (Prentice-Hall, Upper Saddle River, NJ, 1996).

[2] T. D. Gillespie, *Fundamentals of Vehicle Dynamics* (Warrendale, Pennsylvania, Society of Automotive Engineers, 1992).

[3] M. Denny, "The dynamics of antilock brake systems," Eur. J. Phys. **26**, 1007–1016 (2005).

[4] H. Bauer (Editor), *Bosch Automotive Handbook*, 4th ed. (Stuttgart, Robert Bosch GmbH, 1996).

[5] https://everything2.com/title/Stopping+distance

[6] http://www.transport.qld.gov.au/Home/Safety/Road/Speeding/Speeding_stopping_distances

2

Simple Thermodynamics of Jet Engines

P. Patrício and J. M. Tavares

CONTENTS

We use the first and second laws of thermodynamics to analyse the behavior of an ideal jet engine (reversible and with no losses). Simple analytical expressions for the thermal efficiency, the overall efficiency and the reduced thrust are derived. We show that the thermal efficiency depends only on the compression ratio, r, and on the velocity of the aircraft, while the other two performance measures depend also on the ratio between the temperature at the turbine and the inlet temperature in the engine, T_3/T_i. A detailed analysis of these expressions shows that it is not posible to define an optimum set of engine's parameters, $r, T_3/T_i$: larger overall efficiencies lead, in some situations, to smaller thrusts (and vice-versa). We conclude that larger values of r are adequate for moderate velocities, but are limited by T_3/T_i. On the contrary, smaller values of r are adequate for large velocities. Finally, we study how irreversibilities in the compressor and in the turbine decrease the overall efficiency of jet engines and show that this effect is more pronounced for smaller T_3/T_i.

Reproduced from P. Patrício and J. M. Tavares, "Simple thermodynamics of jet engines," American Journal of Physics **78**, 809–814 (2010), https://doi.org/10.1119/1.3373924, with the permission of the American Association of Physics Teachers.

DOI: 10.1201/9781003187103-2

2.1 Introduction

In thermodynamic courses, the first and second laws are usually illustrated in closed systems. Classic textbooks on this subject [1–3] refer always to the paradigmatic Carnot cycle, but also to other known closed cycles (e.g., Rankine, Otto, Diesel and Brayton cycles), calculating their ideal performances based on the powerful concept of reversible transformations. When thermodynamics is introduced to Engineering students, it is convenient to apply its principles also to open stationary systems. In fact, many of the machines used to produce energy or to transfer heat (even those which are given as examples of closed cycles), contain fluids that exchange energy with their neighborhood, or whose thermodynamic properties and velocity change. Jet engines, which can be viewed as an assemblage of several of these devices, provide an important and attractive application of thermodynamic laws to open systems. However, introductory thermodynamics textbooks for engineers do not analyse these engines from a general perspective, but instead give some numerical examples for students to solve.

The purpose of this paper is to show that the detailed study of the jet engine in a generic form allows students to understand the simplicity, the power and the beauty of the laws of thermodynamics, as well as of the crucial role they play in the technological development of such engines. We will show that it is possible to understand with basic thermodynamics: (i) what is a ramjet; (ii) what is the role of the turbine and of the compressor in a jet engine; (iii) why increasing the compression ratio and developing turbines able to withstand high temperatures were key stages in the development of jet engines for commercial aircraft; (iv) how is this development affected by the constraints imposed by the second principle.

This paper is organized as follows. First, we will introduce the basic concepts of thrust, and of overall, propulsive and thermal efficiencies, which will help us to measure the performances of jet engines. Next, we will describe the simplest possible jet engine, and analyse its thermal efficiency. We will then consider an engine with a turbine and a compressor, and study the effects of these on thermal efficiency. This will be followed by an analysis of the overall efficiency and of the thrust. These quantities will depend also on the maximum temperature attained, and we discuss the consequences this dependence for finding the best design for a jet engine. We will then investigate the effects of irreversibilities in the compressor and the turbine on the overall efficiency of the engines. Finally, we will draw some conclusions and discuss possible ways to introduce the results of this paper to students.

2.2 Performances of jet engines

The motion of an aircraft is due to air propulsion. As it flies, the aircraft engine accelerates the air, and accordingly to Newton's law, produces a thrust F which is equal to the difference between the exit and inlet flow velocities, respectively C_e and C_i, times the mass airflow rate $\dot{m}$ (the mass of air that is accelerated per unit time) [1]:

$$F = \dot{m}(C_e - C_i). \tag{2.1}$$

To accelerate the air, there are two main types of engines. The piston engine creates mechanical work that is transmitted to a fan (*propeller*), which in turn produces the required thrust. The *jet engine* provides thrust by burning directly the air with fuel [2] in a combustion chamber, and exhausting the high-temperature mixture through a nozzle that accelerates the air. Airplanes commonly combine both types of engine, to profit their best advantages. In this work we will study jet engines only.

To characterize an engine's performance, we define the dimensionless or *reduced thrust* [4]:

$$\bar{F} = \frac{F}{\dot{m}C_s}, \tag{2.2}$$

where C_s is the local speed of sound. Usually, in aircraft the mass airflow rate $\dot{m}$ is associated with the physical size of an engine, including diameter and weight. Thus the reduced thrust represents the trends of the thrust-to-weight ratio.

The *overall efficiency* η of an aircraft engine is defined as the ratio of the (mechanical) propulsive power to the (thermal) power obtained when fuel is burned [4–6]. If we assume that all the energy released in fuel combustion is absorbed by the air (*stoichiometric conditions* [5]) then,

$$\eta = \frac{FC_i}{\dot{m}q}, \tag{2.3}$$

where q is the heat per unit mass absorbed by the air. It is useful to write down the overall efficiency as the product [6]

$$\eta = \eta_{th} \times \eta_p, \tag{2.4}$$

where η_{th} is the *thermal efficiency*—defined as the ratio of the rate of production of kinetic energy to the fuel power –,

$$\eta_{th} = \frac{\Delta e_c}{q}, \tag{2.5}$$

[1] This expression is valid when the inlet air pressure is equal to its exit pressure, an assumption that we will use throughout the article.

[2] We will assume that the mass fuel rate $\dot{m}_f \ll \dot{m}$.

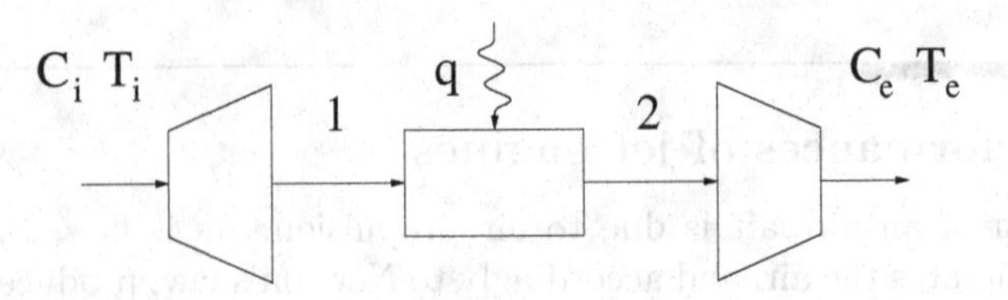

FIGURE 2.1
The simplest model of a jet engine. The air enters the engine through a diffuser, goes to the combustion chamber, and is exhausted through a nozzle.

and η_p is the *propulsive efficiency*—defined as the ratio of propulsive power to the rate of production of kinetic energy –,

$$\eta_p = \frac{FC_i}{\dot{m}\Delta e_c}. \tag{2.6}$$

In the last two equations, Δe_c represents the variation of the kinetic energy of air per unit mass:

$$\Delta e_c = \frac{1}{2}(C_e^2 - C_i^2). \tag{2.7}$$

2.3 The simplest model of a jet engine

Figure 2.1 illustrates the simplest model we can conceive for a jet engine. The air outside is at temperature T_i, and its speed relative to the aircraft, the inlet velocity, is C_i. The air enters the engine through a diffuser, which lowers its speed and increases its pressure. The air then goes into the combustion chamber, and each unit mass of air absorbs the heat q, increasing its internal energy. Finally, the air is exhausted through a nozzle, which accelerates the air until it attains the exit velocity C_e, at temperature T_e.

To analyse this engine, we will apply to each of the three components described in Fig. 2.1—the diffuser, the combustion chamber and the nozzle—the same equation:

$$q - w = \Delta h + \Delta e_c, \tag{2.8}$$

which expresses the conservation of energy for open stationary systems with a constant mass flow rate, and should be familiar to all engineering students with at least one semester of thermodynamics. Here, w is the work per unit mass performed by the air, Δh is the difference between the exit and inlet specific enthalpies, and $\Delta e_c = \Delta C^2/2$ is the difference between the exit and inlet kinetic energies per unit mass. Usually, this equation also comprises another term accounting for the potential energy, which we will simply neglect here.

If we apply this equation to the jet engine consisting of the above three components, and assume that the air has a constant-pressure specific heat c_P, the kinetic energy increase is simply given by:

$$\Delta e_c = \frac{1}{2}(C_e^2 - C_i^2) = q - \Delta h = q - c_P(T_e - T_i). \tag{2.9}$$

This means that if the exit temperature is the same as the initial external temperature, all the heat that is given to the system in the combustion chamber will be used to increase the kinetic energy of the air, leading to maximum thrust.

However, it is known by the second law of thermodynamics that this complete conversion of heat into work (or into kinetic energy of the air) is impossible, and that the most favorable situation (maximum production or minimum consumption of work) is realised if all the processes are reversible. In the subsequent analysis we will assume that all the transformations experienced by the air in the jet engine are reversible. The adiabatic and reversible transformation of the air in the nozzle leads to the relation:

$$T_e = T_2 \left(\frac{P_2}{P_e}\right)^{\frac{1-\gamma}{\gamma}}, \tag{2.10}$$

where P_e is the atmospheric pressure outside (which then equals P_i), T_2 and P_2 are the temperature and pressure of the air upstream of the nozzle (see Fig. 2.1) and γ is the adiabatic coefficient of air, which we assume constant, to simplify our calculations. A similar relation may be found for the adiabatic reversible diffuser:

$$T_i = T_1 \left(\frac{P_1}{P_i}\right)^{\frac{1-\gamma}{\gamma}}. \tag{2.11}$$

If we assume that air pressure is approximately constant in the combustion chamber, $P_1 = P_2$, we may then combine Eqs. (2.10) and (2.11) to obtain:

$$T_e = \frac{T_2}{T_1} T_i. \tag{2.12}$$

Now, the temperatures T_1 and T_2 may be calculated as a function of T_i from the conservation of energy (Eq. (2.8)) applied to the diffuser and to the combustion chamber. Neglecting the relative velocities of the air inside the engine, the temperature after the diffuser is

$$T_1 = T_i + \frac{1}{2}\frac{C_i^2}{c_P} = T_i(1+\epsilon), \tag{2.13}$$

where ϵ is an important non-dimensional parameter for this system,

$$\epsilon = \frac{C_i^2}{2c_P T_i}. \tag{2.14}$$

This reduced kinetic energy [5] of the aircraft is related to the Mach number through

$$\mathcal{M} = \frac{C_i}{C_s} = \frac{C_i}{\sqrt{\gamma r_g T_i}} = \sqrt{\frac{2c_V}{r_g}}\sqrt{\epsilon} \approx 2.24\sqrt{\epsilon}, \tag{2.15}$$

where $C_s = \sqrt{\gamma r_g T_i}$ is the speed of sound in the air at temperature T_i, r_g is the air gas constant, and c_V is the air specific heat at constant volume. Here, we have used the well-known thermodynamical relations $r_g = c_P - c_V$ and $\gamma = c_P/c_V$. We have also used the numerical values $r_g = 287\text{J}/(\text{kgK})$ and $c_V = 718\text{J}/(\text{kgK})$ (which give $\gamma = 1.4$), valid for the air at $T_i = 300$ K. Finally, the temperature after the combustion chamber is related to T_1 by:

$$T_2 = T_1 + \frac{q}{c_P}. \tag{2.16}$$

If we insert all these expressions for the temperatures into Eq. (2.9), we obtain a very simple result for the thermal efficiency of the simplest jet engine:

$$\eta_{th} = \frac{\Delta e_c}{q} = 1 - \frac{1}{1+\epsilon} = \frac{\epsilon}{1+\epsilon} \tag{2.17}$$

This efficiency depends only on $\mathcal{M}$ and is plotted in Fig. 2.2 (black line). We can see that there is no increase in kinetic energy for an aircraft at rest ($\mathcal{M} = 0$). Therefore, a propeller would be needed to start an aircraft equipped with this simplest of jet engines. For small ϵ, or velocities below the speed of sound, the thermal efficiency is approximately $\eta_{th} = \epsilon < 20\%$. Only for very high velocities, several times larger than the speed of sound, does the thermal efficiency approach unity.

The simplest model of a jet engine we have described, without compressor and turbine, corresponds to the ideal ramjet.

2.4 Jet engine with an ideal compressor and turbine

To achieve a much higher thermal efficiency for small velocities, one must simply add a compressor before (i.e., upstream) the combustion chamber.The work used to drive the compressor is generated by a turbine placed just after (i.e., downstream) the combustion chamber, as depicted in Fig. 2.3. When heat losses in the compressor and the turbine are neglected, this work is related to the temperature differences upstream and downstream these devices:

$$w = c_P(T_4 - T_3) = -c_P(T_2 - T_1). \tag{2.18}$$

Since the transformations undergone by air in the compressor and the turbine are also assumed adiabatic and reversible,

$$T_4 = T_3\left(\frac{P_4}{P_3}\right)^{\frac{1-\gamma}{\gamma}}, \qquad T_1 = T_2\left(\frac{P_2}{P_1}\right)^{\frac{1-\gamma}{\gamma}}. \tag{2.19}$$

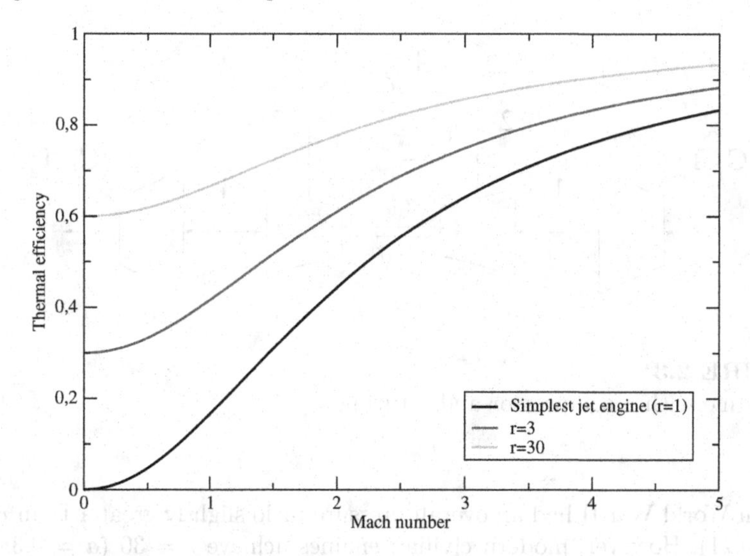

FIGURE 2.2
Thermal efficiency η_{th} of a jet engine, as a function of its velocity $\mathcal{M} = C_i/C_s$ (Mach number), for the simplest jet engine, with no compressor, and for pressure ratios $r \approx 3$ ($a = 0.7$) and $r \approx 30$ ($a = 0.4$).

The compression ratio $r = P_2/P_1$ depends on the manufacturing characteristics of the compressor. Using the results obtained in the previous section, the thermal efficiency of the new jet engine can now be written:

$$\eta_{th} = \frac{\Delta e_c}{q} = 1 - \frac{a}{1+\epsilon} \tag{2.20}$$

where $0 < a = r^{\frac{1-\gamma}{\gamma}} \leq 1$, since both the compression ratio r and the adiabatic coefficient γ are greater than one. Although the engine consists of five different components, the final result for its thermal efficiency, Eq. (2.20), is amazingly simple: it depends only on a (or r) and ϵ and is easy to analyse in a thermodynamics class. [3] If $r = a = 1$, there is no compression at all, and we obtain the thermal efficiency of the simplest jet engine (see Eq. (2.17)). Early jet engines

[3]It is surprising that this result is hardly ever mentioned in classical textbooks. In fact, the thermal efficiency of a jet engine is sometimes given as an example of application of the Brayton cycle, which is composed by two isentropic and two isobaric transformations. This closed cycle can in fact mimic the thermodynamics of airflow in the jet engine: an isentropic compression from the inlet to the combustor, an isobaric heating in the combustion chamber, an isentropic expansion in the turbine and in the nozzle, and an (artificial) isobaric cooling of the air back to the inlet. It is also common to express the efficiency of this cycle in terms of the ratio of the pressure at the inlet P_i to the pressure in the combustion chamber. Consequently, this analysis does not split the increase in pressure into contributions from the inlet (dependent on C_i or ϵ) and from the compressor (dependent on r or a). Moreover, as in the usual analysis of closed cycles, the aim is to produce work: but we know that, in

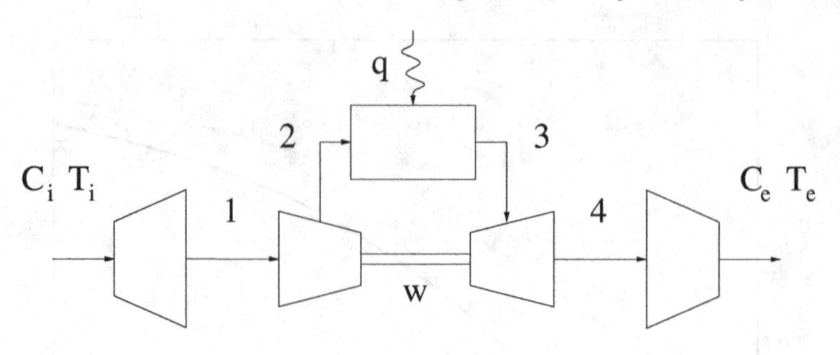

FIGURE 2.3
Jet engine with a compressor and a turbine

used in World War II had an overall pressure ratio slightly greater than $r = 3$ ($a = 0.71$). However, modern civilian engines achieve $r = 30$ ($a = 0.38$, for the Boeing 747) or even $r = 40$ ($a = 0.35$, for the Airbus A380), which for regulatory reasons do not exceed the speed of sound. It is also instructive to know that Concorde's engine has $r = 15$ ($a = 0.46$).

For comparison, we present the thermal efficiencies of these jet engines in Fig. 2.2, for $a = 0.7$ ($r \approx 3$, in red) and $a = 0.4$ ($r \approx 30$, in green). For small ϵ, we may write $\eta_{th} = (1 - a) + a\epsilon$. At rest, the thermal efficiency substantially improves as r is increased (this is illustrated for $\eta_{th} = 30\%$ and $\eta_{th} = 60\%$). The initial slope of η_{th} gets smaller as the pressure ratio increases. Nevertheless, these lines only join at infinite speeds, for which we have $\eta_{th} = 1$.

The results obtained for the thermal efficiency give a first indication of the qualitative behavior of jet engines. In particular, the thermal efficiency is a very simple function of the speed of the aircraft and of the compression ratio of the compressor only—no dependence on the absorbed heat was found. The thermal efficiency is the same, whether we use a small or a large amount of heat: the kinetic energy of the aircraft varies proportionally. According to these ideal results, if we need to travel faster, we should just use more fuel (to increase the heat). But there is one important consequence: the temperature inside the jet engine will be higher and higher. At some point, we risk melting the compressor or the turbine.

---

jet engines, the aim is to produce thrust. Therefore, direct application of the analysis of the Brayton cycle to the jet engine yields a very limited description.

2.5 Overall efficiency and thrust

Besides the compression ratio, the inlet temperature of the turbine T_3 (see Fig. 2.3), which is the highest temperature reached by air inside the engine, is the second most important manufacturing parameter of the jet engine that we will take into account. This temperature is easily related to the absorbed heat using the first law of thermodynamics:

$$\tilde{T}_q = \frac{q}{c_P T_i} = \frac{T_3}{T_i} - \frac{1+\epsilon}{a}, \tag{2.21}$$

where we have introduced the dimensionless absorbed heat temperature $\tilde{T}_q$. Using the definition of ϵ and Eq. (2.7), the propulsive efficiency can be related to η_{th} and $\tilde{T}_q$:

$$\eta_p = \frac{2}{1+\sqrt{1+\frac{\eta_{th}\tilde{T}_q}{\epsilon}}}. \tag{2.22}$$

The overall efficiency $\eta = \eta_{th} \times \eta_p$ and the reduced thrust

$$\bar{F} = \frac{F}{\dot{m}C_s} = 2.24\sqrt{\epsilon}\left[-1+\sqrt{1+\frac{\eta_{th}\tilde{T}_q}{\epsilon}}\right], \tag{2.23}$$

depend not only on ϵ and r (like η_{th}), but also on T_3/T_i, through $\tilde{T}_q$ (see Eq. (2.21)).

It is interesting to verify that the condition $F > 0$ (or $q > 0$) imposes an upper bound for the thermal efficiency,

$$\eta_{th} < 1 - \frac{T_i}{T_3}, \tag{2.24}$$

which is nothing more than Carnot's theorem. Consequently, given a pair of the engine design parameters, r and T_3/T_i, there is a velocity ϵ_{max} beyond which flight becomes impossible. This velocity is obtained by setting $\eta_{th} = 1 - T_i/T_3$ (or $F = 0$, or $q = 0$),

$$\epsilon_{max} = aT_3/T_i - 1. \tag{2.25}$$

For a fixed T_3/T_i, the increase in r (decrease of a) will always lower the maximum possible velocity; if $a < T_i/T_3$ the jet engine will not even work. Therefore, we may only have engines with large compression ratios if, at the same time, the turbine materials can withstand high enough T_3.

The analysis of the interplay between the two design parameters and its effects on the overall efficiency and reduced thrust are shown in Figs. 2.4 and 2.5, where we plot, respectively, η (from Eqs. (2.20) and (2.22)) and Eq. (2.23) as a function of $\mathcal{M}$, for several fixed values of r and T_3/T_i.

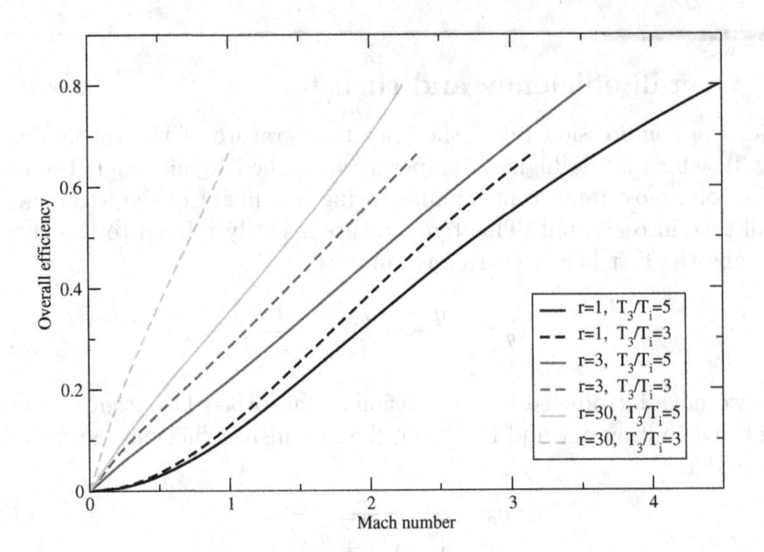

FIGURE 2.4
Overall efficiency η of a jet engine as a function of its velocity $\mathcal{M}$, for the values of a and T_3/T_i given in the inset.

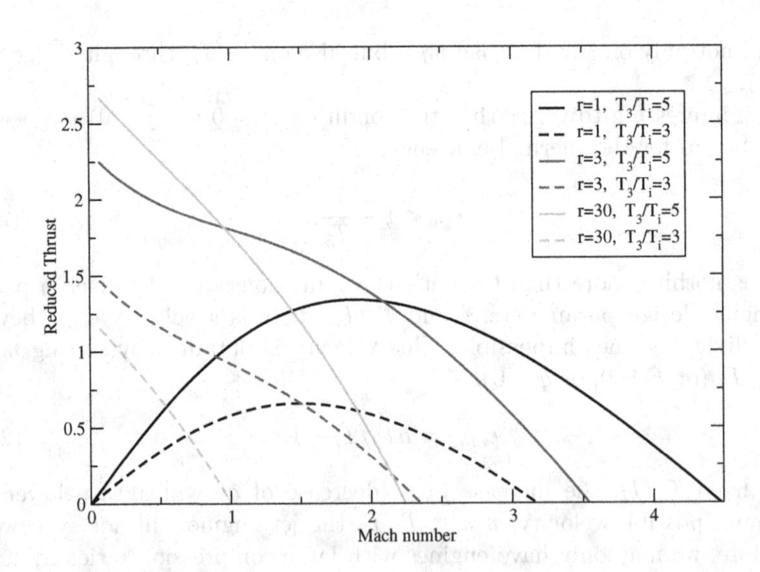

FIGURE 2.5
Reduced thrust $\bar{F}$ as a function of $\mathcal{M}$, for the values of a and T_3/T_i indicated in the inset.

The results of Fig. 2.4 confirm that increasing r (at fixed T_3 and $\mathcal{M}$) increases the efficiency, but only up to a certain limit (the value of r for which $\epsilon = \epsilon_{max}$). It also shows that: (a) at all velocities and for fixed r, an increase in T_3/T_i decreases η; (b) for fixed design parameters r and T_3/T_i, increasing $\mathcal{M}$ increases η. If $T_3/T_i \approx 5$ is assumed, this figure shows why Concorde (which flew at $\mathcal{M} \approx 2$) and was designed with $r \approx 15$ ($a \approx 0.45$) had higher efficiencies than other commercial and larger airplanes (with $\mathcal{M} \approx 0.75$ and $r \approx 30$ ($a \approx 0.4$)).

The results presented in Fig. 2.5 show that reduced thrust decreases with increasing velocity (for a fixed pair $(T_3/T_i, r)$), except for the ramjet ($r = 1$), for which there is a velocity that maximizes the thrust and some auxiliary device is needed to start the motion (since $\bar{F} = 0$ when $\mathcal{M} = 0$). An increase in T_3/T_i increases the reduced thrust, for every pair $(\mathcal{M}, r)$. The dependence of the thrust on r for fixed $(T_3/T_i, \mathcal{M})$ is more complex. If we want to use the curves in Fig. 2.5 to guess the value of r that maximizes the thrust, we have to consider the following cases:

(a) low velocities ($\mathcal{M} < 1$): if $T_3/T_i = 3$, the maximum thrust is obtained for $r = 3$; but if $T_3/T_i = 5$ the maximum is for $r = 30$;

(b) intermediate velocities ($1 < \mathcal{M} < 2$): if $T_3/T_i = 5$ the maximum thrust is now obtained for $r = 3$; for $T_3/T_i = 3$, the maximum thrust is obtained for $r = 3$ up to $\mathcal{M} \approx 1.5$; and for $r = 1$ for $\mathcal{M} > 1.5$.

(c) high velocities ($\mathcal{M} > 2$): the maximum thrust is always obtained for the ramjet $r = 1$.

The combined analysis of the results of these two figures shows that the problem of finding the best r of an engine for a plane that travels at a given velocity is non-optimal, owing to the upper bound on the temperature T_3/T_i at the turbine. In fact, one would like to maximize both reduced thrust and efficiency (to reduce fuel consumption). Figure 2.4 shows that, to increase the efficiency, r should be increased up to its maximum possible value. On the other hand, Fig. 2.5 reveals that maximization of reduced thrust with a particular value of r depends on $\mathcal{M}$ and T_3/T_i.

In conclusion, by applying the laws of thermodynamics to this model of the jet engine we have shown that it is not possible to maximize reduced thrust and efficiency at the same time. Therefore, the specific flight requirements of the aircraft—whether we want a fast military aircraft or an efficient commercial plane—will determine the best choice of parameters.

2.6 Non-ideal compressor and turbine

Ideal jet engines do not exist: some heat is lost in all its components and, more interestingly, even if these losses are neglected, irreversibilities will affect engine performance. Suppose, for example, that almost no heat is supplied to this system. In this case, how could the compressor use the work generated by the turbine to compress the air exactly in the same way as if more heat were available?

In this section we consider a more realistic compressor and turbine: the effect of irreversibilities is introduced through the isentropic efficiencies, respectively η_c and η_t, which compare the work consumed or produced in an adiabatic irreversible process to that involved in an ideal (reversible) process:

$$\eta_c = \frac{w_s}{w_r} = \frac{T_{2s} - T_1}{T_{2r} - T_1}, \qquad \eta_t = \frac{w_r}{w_s} = \frac{T_{4r} - T_3}{T_{4s} - T_3}. \tag{2.26}$$

The subscripts s and r denote the ideal (isentropic) and real transformations, and we have used energy conservation (Eq. (2.8)). These efficiencies are typically of the order of 70 to 90%.

Irreversibilities will directly affect the thermal efficiency η_{th}, which will in turn affect the propulsive efficiency and the reduced thrust (see Eqs. (2.22) and (2.23))—$\tilde{T}_q$ remains the same.

If we introduce these new parameters, and re-calculate the final exit temperature, we get a complicated expression:

$$\frac{T_e}{T_i} = \frac{a(\tilde{T}_q+(1+\epsilon))((1+\epsilon)(1+a(\eta_c-1))+a\tilde{T}_q\eta_c)\eta_t}{(1+\epsilon)(a\tilde{T}_q\eta_c\eta_t+(1+\epsilon)(\eta_t-1+a(1+(\eta_c-1)\eta_t)))}. \tag{2.27}$$

We can see why most (if not all) textbooks on thermodynamics do not use this particular result, and instead present the non-ideal jet engine as a numerical example.

To obtain the final thermal efficiency, we just have to use this expression for T_e/T_i in Eqs. (2.9) and (2.5). η_{th} will depend not only on the compression ratio r (or a) and the reduced kinetic energy ϵ, but also on the isentropic efficiencies η_c and η_t. More interestingly, and unlike in the ideal case, the thermal efficiency of the irreversible jet engine will now depend non-trivially on the absorbed heat q (or $\tilde{T}_q$).

Nevertheless, some simple results can be derived for limiting situations. If the absorbed heat is very large ($T_q \gg T_i$), we recover the ideal thermal efficiency:

$$\lim_{q\to\infty} \eta_{th} = 1 - \frac{a}{1+\epsilon}. \tag{2.28}$$

However, as the absorbed heat decreases, so does the thermal efficiency. Eventually, if the absorbed heat is sufficiently small, the work generated by the irreversible turbine is no longer able to activate the compressor, and the whole engine does not work (unless the air decelerates).

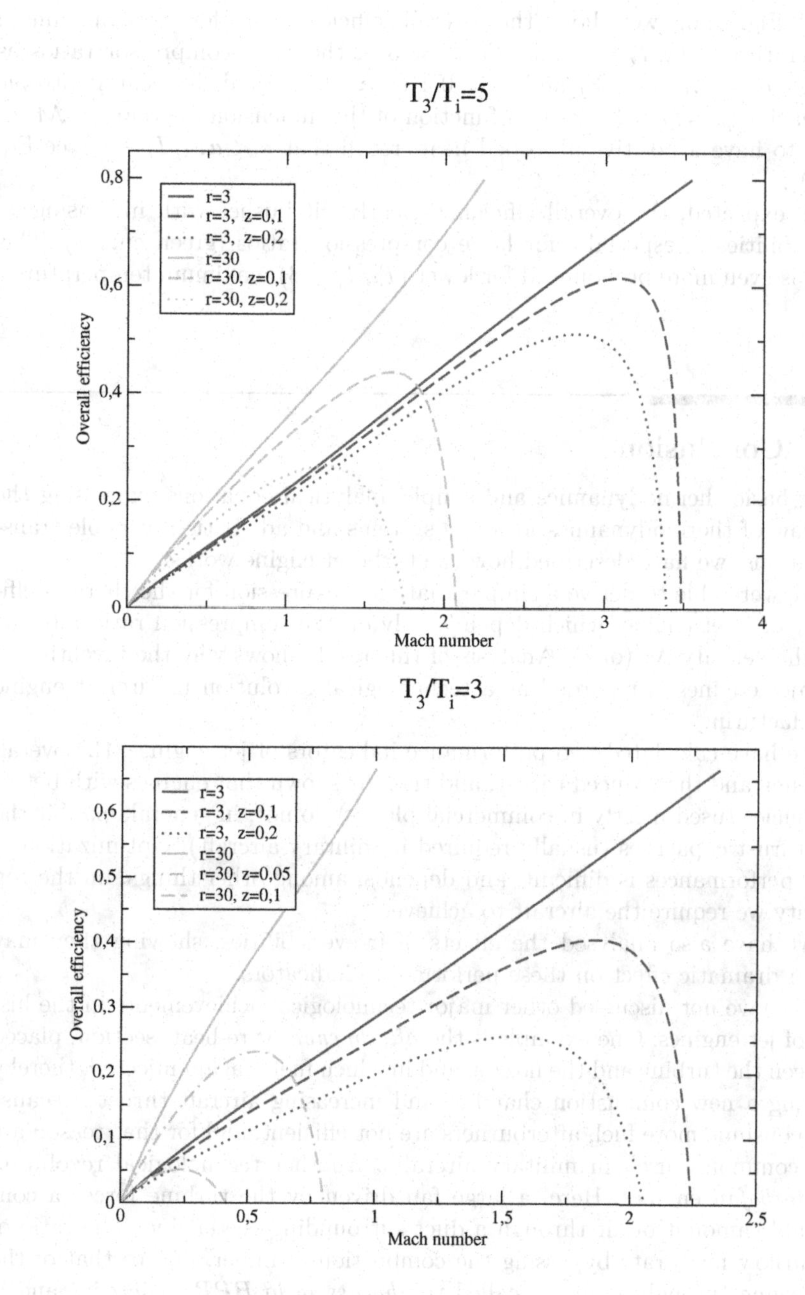

FIGURE 2.6
Constrained efficiencies of the jet engine.

In Fig. 2.6, we show the overall efficiency η, for two maximum temperatures—$T_3/T_i = 5$ and $T_3/T_i = 3$ –, the same compressor ratios as before—$a = 0.7$ ($r \approx 3$) and $a = 0.4$ ($r \approx 30$) –, and different isentropic efficiencies $\eta_c = \eta_t = 1 - z$, as a function of the dimensionless velocity $\mathcal{M}$. In order to have a positive absorbed heat, recall that $\epsilon < aT_3/T_i - 1$ (see Eq. (2.25)).

As expected, the overall efficiency clearly diminishes with increasing irreversibilities z, especially for large compression ratios (green curves). This effect is even more pronounced for lower ($T_3/T_i = 3$) maximum temperatures.

2.7 Conclusion

Using basic thermodynamics and simple analytical equations expressing the first law of thermodynamics for open systems and adiabatic reversible transformations, we have described how the turbojet engine works.

We were able to derive a simple analytical expression for the thermal efficiency of a jet engine, which depends only on the compression ratio r (or a) and the velocity $\mathcal{M}$ (or ϵ). Analysis of this result shows why the invention of turbojet engines is regarded as a technological revolution in aircraft engine manufacturing.

We have calculated two performance indicators of jet engines, the overall efficiency and the reduced thrust, and thereby shown that engines with better efficiencies (used mostly in commercial planes) do not, in general, exhibit the best thrust capacities (usually required in military aircraft). Optimization of these performances is difficult, and depends, among other things, on the top velocity we require the aircraft to achieve.

We have also analysed the effects of irreversibilities, showing they may have a dramatic effect on these performance indicators.

We have not discussed other major technological achievements in the history of jet engines. One example is the *afterburner*, or re-heat, section, placed between the turbine and the nozzle, and in which fuel is again injected, thereby creating a new combustion chamber and increasing aircraft thrust. Because they consume more fuel, afterburners are not efficient, and for that reason are only commonly used in military aircrafts. Another technological revolution was *turbofan* engines. Here, a large fan driven by the turbine forces a considerable amount of air through a duct surrounding the engine. The ratio of the airflow mass rate bypassing the combustion chamber, $\dot{m}_d$, to that of the air flowing through it, $\dot{m}_c$, is called the *bypass ratio* $BPR = \dot{m}_d/\dot{m}_c$, and is typically around $5 - 6$. Turbofans reduce fuel consumption considerably, and are responsible for the success of jumbo planes, which carry a few hundred people at speeds of almost 1000 km/h. New engines, called *propjets*, achieve still higher efficiencies, with bypass ratios of the order of 100.

Our study described the basic thermodynamic concepts of jet engines, presenting results that are easy to explain in an introductory engineering course. The detailed analysis of jet engines is obviously complex, and covered in a huge number of scientific books—from the description of engines to the aerodynamical or structural problems associated with aircraft ([7,8], to name just a few examples). Still, the numerical analysis presented in this paper can easily be extended by students to some other interesting problems, such as: finding, for fixed $\mathcal{M}$ and T_3/T_i, which value of r maximizes thrust; calculating the effects of irreversibilities on thermal efficiency and reduced thrust, etc. We hope that our article may arouse the curiosity of readers and encourage them to undertake more profound studies of this fascinating subject.

Acknowledgments

The authors wish to thank P. I. C. Teixeira for a careful reading of the manuscript.

References

[1] P. M. Fishbane, S. Gasiorowicz, and S. T. Thornton, *Physics for Scientists and Engineers*, 2nd ed. (Prentice-Hall, Upper Saddle River, NJ, 1996).

[2] A. Cengel and M. A. Boles, *Thermodynamics: An Engineering Approach*, 3rd ed. (McGraw-Hill, New York, 1998).

[3] M. J. Moran and H. N. Shapiro, *Fundamentals of Engineering Thermodynamics*, 5th ed. (John Wiley and Sons, Hoboken, NJ, 2006).

[4] R. D. Flack, *Fundamentals of Jet Propulsion with Applications* (Cambridge University Press, New York, 2005).

[5] W. H. Heiser and D. T. Pratt, *Hypersonic Airbreathing Propulsion* (AIAA, Reston, VA, 1994).

[6] Z. S. Spakowszky, "Thermodynamics and propulsion," (web.mit.edu/16.unified/www/SPRING/propulsion/notes/notes.html).

[7] J. D. Anderson, Jr., *Fundamentals of Aerodynamics*, 2nd ed. (McGraw-Hill, Singapore, 1991).

[8] T. H. G. Megson, *Aircraft Structures for Engineering Students*, 4th ed. (Elsevier, Oxford, 2008).

3

Surprises of the Transformer as a Coupled Oscillator System

J. P. Silva and A. J. Silvestre

CONTENTS

We study a system of two Resistance, Inductance, Capacity (RLC) oscillators coupled through a variable mutual inductance. The system is interesting because it exhibits some peculiar features of coupled oscillators: i) there are two natural frequencies; ii) in general, the resonant frequencies do not coincide with the natural frequencies; iii) the resonant frequencies of both oscillators differ; iv) for certain choices of parameters, there is only one resonant frequency, instead of the two expected.

3.1 Introduction

One of us (JPS) teaches a course on "Vibrations and Waves". Early in the course, students study the RLC system and learn that:

1. If there were no resistance, the system would have a natural frequency $\omega_0 = 1/\sqrt{LC}$.

Reproduced from J. P. Silva and A. J. Silvestre, "Surprises of the transformer as a coupled oscillator system," European Journal of Physics **29**, 413–420 (2008), https://doi.org/10.1088/0143-0807/29/3/002, with the permission of IOP Publishing Ltd.

DOI: 10.1201/9781003187103-3

2. In the underdamped regime, the natural frequency is $\sqrt{\omega_0^2 - \gamma^2/4}$, with $\gamma = R/L$—this occurs for $\gamma < 2\omega_0$.

3. When the system is driven with an external power supply, $\text{Re}\left\{V_0\, e^{i\,\omega_F\, t}\right\}$, the steady state solution for the charge is given by

$$q(t) = \text{Re}\left\{Q_F\, e^{-i\,\delta_F}\, e^{i\,\omega_F\, t}\right\}, \tag{3.1}$$

where

$$\begin{aligned} Q_F &= \frac{V_0/L}{\sqrt{(\omega_0^2 - \omega_F^2)^2 + \gamma^2\omega_F^2}}, \\ \tan\delta_F &= \frac{\gamma\omega_F}{\omega_0^2 - \omega_F^2}. \end{aligned} \tag{3.2}$$

4. In the underdamped regime, $q(t)$ and, thus, the voltage across the capacitor in the steady state has its maximum for $\omega_F = \sqrt{\omega_0^2 - \gamma^2/2}$ (if $\gamma < \sqrt{2}\omega_0$) or for $\omega_F = 0$ (if $\sqrt{2}\omega_0 < \gamma < 2\omega_0$).

5. In any regime, the maximum of the steady state current $i(t) = dq(t)/dt$ and, thus, of the voltage across the resistance occurs always for $\omega_F = \omega_0$, regardless of γ. This frequency ω_F of the external power supply for which the steady state current has its maximum amplitude is known as the resonant frequency.

After learning about coupled systems, some students suggested the study of an experimental apparatus which they had seen in the "Electromagnetism" course, consisting of a solenoid which can be slided in and out of a slightly larger solenoid.

Coupled oscillators are discussed in the literature in the context of mechanical systems [1], but their electrical analogues would have capacitive (rather than inductive) coupling. Frank and Brentano [2] study the normal modes of two coupled LC circuits as analogues to quantum mechanical level repulsion, but with no resistance or comparison with the forced regime. A very interesting system of coupled RLC circuits was discussed by Hansen, Harang, and Armstrong [3]. They study the forced regime, but do not compare its resonances with the natural frequencies of the free regime. Some books do compare the natural frequencies with the resonant frequencies, but they do so in simplified contexts where these frequencies coincide, and include explicit or implicit statements promoting the coincidence between resonant and natural frequencies to a general rule [4].

Using what they had learned about the current on the RLC oscillator our students made the following prediction. First, that there would be two natural frequencies, because there are two oscillators. Second, that the steady state resonant frequencies would coincide with the natural frequencies, because that was the case with only one RLC oscillator. Third, that the resonant frequencies of both oscillators would be the same, because they had the same components.

And fourth, that there would always be two resonant frequencies, because there are two oscillators. This is a very interesting exercise because the last three "predictions" turn out to be false. The correct comparison between natural and resonant frequencies in a system of coupled RLC circuits and the surprises that it entails constitute the main thrust of our work.

In Sec. 3.2, we set up the problem and determine the natural frequencies of the system. In Sec. 3.3, we study the driven system and determine the resonant frequencies for the currents in each oscillator. We draw our conclusions in Sec. 3.4.

3.2 Natural frequencies of a transformer

Let us consider a circuit with a resistor R, a capacitor C, and a solenoid of inductance L, which can be slided in and out of a second solenoid with the same inductance. In turn, the second solenoid forms a circuit with identical resistor R and capacitor C [1]. As the relative position of the two solenoids is altered, the mutual inductance M changes [2]. The system is shown schematically in Fig. 3.1.

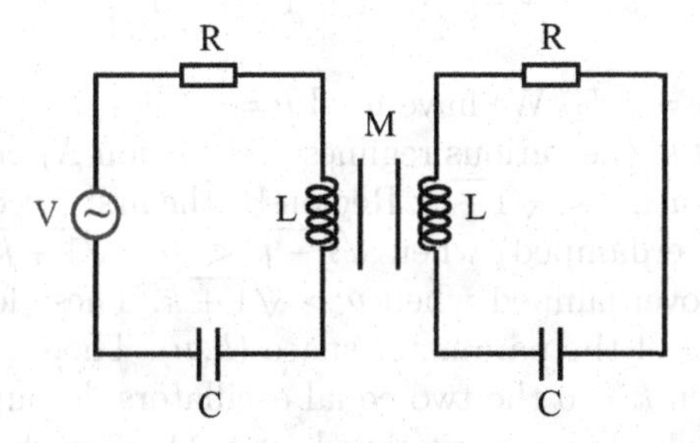

FIGURE 3.1
Two RLC circuits connected through a mutual inductance M. In Sec. 3.2 (Sec. 3.3), we consider this circuit without (with) the external power source V.

[1] We take the two RLC circuits to be identical. Of course, this is difficult to achieve if one solenoid is to be inserted inside the other. We do this because the calculations are considerable simplified, while retaining most of the relevant physics we wish to highlight.

[2] Without loss of generality, we will take M and $k = M/L$ positive, and $k \to 1$ when one solenoid surrounds completely the other.

The circuit equations are

$$\begin{aligned} \frac{q_1}{C} + R\, i_1 + L\frac{d\, i_1}{dt} + M\frac{d\, i_2}{dt} &= 0, \\ \frac{q_2}{C} + R\, i_2 + L\frac{d\, i_2}{dt} + M\frac{d\, i_1}{dt} &= 0. \end{aligned} \tag{3.3}$$

Substituting for complex normal mode solutions ($\alpha = 1, 2$),

$$q_\alpha(t) = \frac{\hat{I}_\alpha}{i\omega}\, e^{i\,\omega\, t}, \tag{3.4}$$

we obtain

$$\begin{aligned} (\omega_0^2 - \omega^2 + i\gamma\omega)\, \hat{I}_1 - k\omega^2 \hat{I}_2 &= 0, \\ -k\omega^2 \hat{I}_1 + (\omega_0^2 - \omega^2 + i\gamma\omega)\, \hat{I}_2 &= 0, \end{aligned} \tag{3.5}$$

where $k = M/L$. Nontrivial solutions require the determinant of the system to vanish, leading to

$$\omega = \frac{\omega_0}{1+k}\left[i\,\nu \pm \sqrt{1+k-\nu^2}\right], \tag{3.6}$$

corresponding to $\hat{I}_1 = \hat{I}_2$, or to

$$\omega = \frac{\omega_0}{1-k}\left[i\,\nu \pm \sqrt{1-k-\nu^2}\right], \tag{3.7}$$

corresponding to $\hat{I}_1 = -\hat{I}_2$. We have used $\nu = \gamma/(2\omega_0)$.

Given a value of k, the various regimes are: Region A) both normal modes are underdamped when $\nu < \sqrt{1-k}$; Region B) the first (second) normal mode is underdamped (overdamped) when $\sqrt{1-k} < \nu < \sqrt{1+k}$; Region C) both normal modes are overdamped when $\nu > \sqrt{1+k}$. These features are shown in Fig. 3.2. in terms of the parameter space (k, ν). There are two interesting limiting cases. When $k = 0$ the two equal oscillators decouple; they are both underdamped ($\nu < 1$), both overdamped ($\nu > 1$), or both critically damped ($\nu = 1$). When $k = 1$ there is maximal coupling between the two systems; the normal mode of Eq. (3.7) is always overdamped, while the normal mode of Eq. (3.6) is underdamped when $\nu < \sqrt{2}$.

We will simplify the discussion in the next section by concentrating mainly on cases for which both normal modes are underdamped. In that case, the first normal mode has

$$i_1(t) = i_2(t) \propto e^{-\frac{\gamma}{2(1+k)}t} \cos(\omega_+\, t), \tag{3.8}$$

and the second normal mode has

$$i_1(t) = -i_2(t) \propto e^{-\frac{\gamma}{2(1-k)}t} \cos(\omega_-\, t), \tag{3.9}$$

with

$$\omega_\pm = \frac{\omega_0}{1 \pm k}\sqrt{1 \pm k - \nu^2}. \tag{3.10}$$

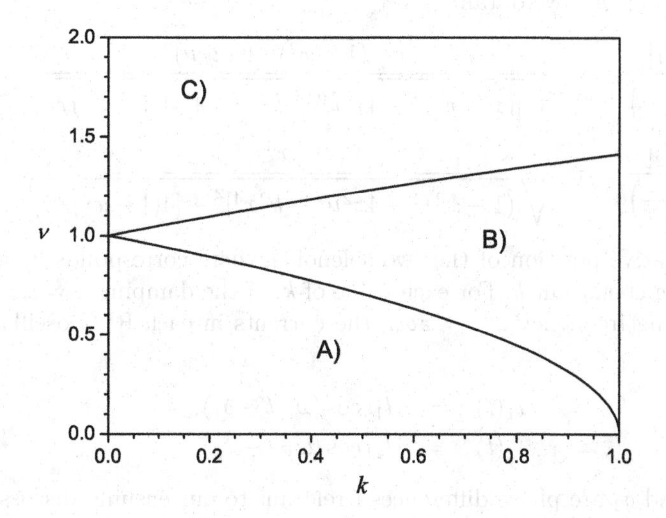

FIGURE 3.2
Damping regimes of the two normal modes in terms of the parameters k and ν. The meaning of regions A), B) and C) is explained in the text.

3.3 Resonant frequencies of a driven transformer

Let us now consider a situation in which the RLC oscillator denoted by 1 is driven by an external power source of frequency ω_F and amplitude V_0. The circuit equations become

$$\begin{aligned} \frac{q_1}{C} + R\, i_1 + L\frac{d\, i_1}{dt} + M\frac{d\, i_2}{dt} &= V_0\, e^{i\,\omega_F\, t}, \\ \frac{q_2}{C} + R\, i_2 + L\frac{d\, i_2}{dt} + M\frac{d\, i_1}{dt} &= 0. \end{aligned} \tag{3.11}$$

We use complex notation, define $x = \omega_F/\omega_0$, and search for steady state solutions ($\alpha = 1, 2$),

$$q_\alpha(t) = \frac{\hat{I}_\alpha}{i\omega_F}\, e^{i\,\omega_F\, t}, \tag{3.12}$$

After some algebra, we obtain

$$\begin{aligned}
\frac{|\hat{I}_1|}{\left(\frac{V_0}{\omega_0 L}\right)} &= \frac{x\sqrt{(1-x^2)^2+(2x\nu)^2}}{\sqrt{\left[(1-x^2)^2-4x^2\nu^2-k^2x^4\right]^2+\left[4(1-x^2)x\nu\right]^2}}, \\
\frac{|\hat{I}_2|}{\left(\frac{V_0}{\omega_0 L}\right)} &= \frac{kx^3}{\sqrt{\left[(1-x^2)^2-4x^2\nu^2-k^2x^4\right]^2+\left[4(1-x^2)x\nu\right]^2}}.
\end{aligned} \tag{3.13}$$

To each relative position of the two solenoids, there corresponds a value for the coupling coefficient k. For each value of k, of the damping $\gamma = 2\omega_0\,\nu$, and of the driving frequency $\omega_F = x\,\omega_0$, the currents in each RLC oscillator are given by

$$\begin{aligned}
i_1(t) &= |\hat{I}_1|\cos(\omega_F\,t - \delta_1), \\
i_2(t) &= |\hat{I}_2|\cos(\omega_F\,t - \delta_2),
\end{aligned} \tag{3.14}$$

where δ_1 and δ_2 are phase differences irrelevant to our ensuing discussions.

As one changes the driving frequency, the amplitudes $|\hat{I}_1|$ and $|\hat{I}_2|$ vary. Figure 3.3 shows this variation, for ($k = 0.5,\ \nu = 0.1$) and for ($k = 0.5,\ \nu = 0.3$). Figure 3.3(a) has two relevant features. The two amplitudes have each

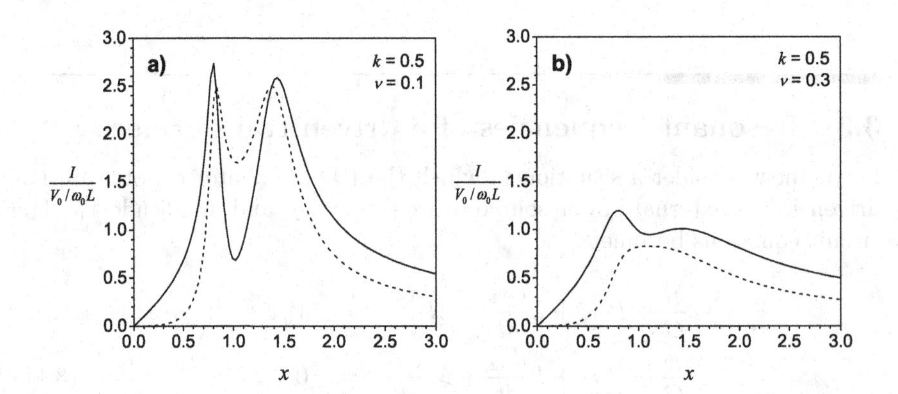

FIGURE 3.3
Variation of the current amplitudes with the driving frequency. The amplitudes $|\hat{I}_1|$ and $|\hat{I}_2|$ are represented by solid and dashed lines, respectively. Figure 3(a) has $k = 0.5$ and $\nu = 0.1$. Figure 3(b) has $k = 0.5$ and $\nu = 0.3$.

two resonances, but these occur for values of x which, although close to each other, are not equal. Our students felt vindicated by the former and slightly worried by the later. They became very worried when Fig. 3(b) showed that $|\hat{I}_2|$ might have only one maximum. Even worse, for ν larger than about 0.4, both $|\hat{I}_1|$ and $|\hat{I}_2|$ show only one resonance.

The values of the normalized driving frequency (x) for which resonances occur can be displayed in terms of the normalized damping (ν), for various choices of the coupling coefficient k. This is shown in Figs. 3.4 for $k = 0.02$, $k = 0.2$, $k = 0.5$, and $k = 0.98$. Also shown as dotted lines are the normalized frequencies of the normal modes of free oscillations $(\omega_\pm/\omega_0)$. The curve corresponding to ω_- is the one that reaches the horizontal axis for the lowest value of ν. Values of ν to the left of this point correspond to free oscillations (and to transient pieces of the forced oscillations) which are underdamped in both normal mode components.

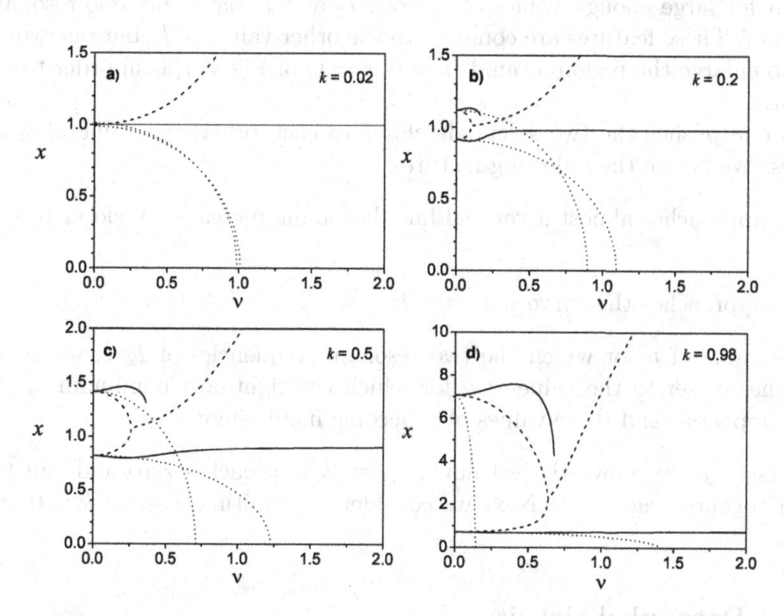

FIGURE 3.4
Values of the normalized driving frequency x, for which resonances occur in the current amplitudes. Resonant frequencies for $\hat{I}_1$ and $\hat{I}_2$ are represented by solid and dashed lines, respectively. Also shown as dotted lines are the normalized frequencies of the normal modes of free oscillations $(\omega_\pm/\omega_0)$. Figures 4(a), 4(b), 4(c), and 4(d) correspond to $k = 0.02$, $k = 0.2$, $k = 0.5$, and $k = 0.98$, respectively. Notice the increase in scale.

In Fig. 3.4(a,b,c, and d) we notice that the resonant frequencies of $\hat{I}_1$ and $\hat{I}_2$ coincide with each other and with the natural frequencies when $\nu = 0$. This can be seen directly by setting $\nu = 0$ in Eqs. (3.13). The denominators on the right-hand sides become $|(1-x^2)^2 - k^2x^4|$, which vanish for $x = 1/\sqrt{1 \pm k} \equiv \omega_\pm/\omega_0$, and lead to divergent currents for these driving frequencies.

Let us concentrate on Fig. 3.4(c). As ν grows, the resonant frequencies of $\hat{I}_1$ and $\hat{I}_2$ become different from the natural frequencies. They also become different from each other, a feature already apparent in Fig. 3.3(a). As ν grows even further, the two resonant frequencies of $\hat{I}_2$ coalesce into only one, a case illustrated by Fig. 3.3(b). Figure 3.3(b) also hints at the possibility that the rightmost maximum of $|\hat{I}_1|$ is slowly disappearing as ν increases further. This is clearly seen as the disappearance on the uppermost solid curve in Fig. 3.4(c). Finally, for large enough values of ν, both $\hat{I}_1$ and $\hat{I}_2$ have only one resonant frequency. These features are common to the other values of k, but one would have to enlarge the region around $(\nu = 0, x = 1)$ of Fig. 3.4(a) in order to see it clearly.

As one pushes the two solenoids closer to each other, the value of k increases. We notice the following features:

- ω_- approaches almost a vertical line, becoming increasingly closer to $\nu = 0$;
- ω_+ approaches the curve $\sqrt{2 - \nu^2}/2$;
- the value of ν for which the two resonant frequencies of $\hat{I}_2$ coalesce becomes closer to the value of ν for which the rightmost maximum of $|\hat{I}_1|$ disappears—and these values of ν become increasingly large.

Conversely, as we move the solenoids apart, k approaches zero and "all the action" occurs near $\nu = 0$. Next we consider the limiting cases of $k = 0$ and $k = 1$.

3.3.1 Decoupled circuits

As the two solenoids are separated, $k \to 0$, the two circuits become decoupled, and the normal frequencies $\omega_\pm$ tend to a common value $\sqrt{\omega_0^2 - \gamma^2/4}$. This can be seen in Fig. 3.4(a) where the dotted curves of $\omega_\pm/\omega_0$ almost overlap. From the introduction, we recognize this common value as the natural frequency of each RLC circuit considered individually.

We also see in Fig. 3.4(a) that, as $k \to 0$, $\hat{I}_1$ will get only one resonant frequency, and that this resonance occurs for $x = 1$, independently of ν. From the introduction, we recognize this as the correct behavior for a single driven RLC circuit.

Figure 3.4(a) is a bit more subtle when it comes to i_2, for which it still seems to show one resonance. The point is that the second Eq. (3.13) leads to $i_2 \to 0$ as $k \to 0$. There really is no current in the second circuit when we drive only the first circuit and decouple the two, as it should be.

3.3.2 Maximum coupling

As the two solenoids are joined (see footnote 1), $k \to 1$, and the two circuits become maximally coupled. We see from Eq. (3.10) that ω_- diverges as $k \to 1$. So, we must go back and set $k = 1$ directly in Eqs. (3.5). Setting the determinant to zero yields two solutions. One solution gives

$$\omega = \frac{\omega_0}{2}\left[i\nu \pm \sqrt{2-\nu^2}\right], \tag{3.15}$$

corresponding to

$$i_1(t) = i_2(t) \propto e^{-\frac{\gamma}{4}t}\cos\left(\frac{\omega_0}{2}\sqrt{2-\nu^2}\,t\right). \tag{3.16}$$

We have assumed that $\nu < \sqrt{2}$ so that this normal mode is underdamped. The other solution gives

$$\omega = i\frac{\omega_0^2}{\gamma}, \tag{3.17}$$

corresponding to $i_1(t) = -i_2(t)$ and a critically damped normal mode. This is consistent with Fig. 3.2.

One may be surprised by the fact that, although both inductors and capacitors are present, there is no oscillation in this case. The reason is very interesting. Setting $M = L$ in Eqs. (3.3) and subtracting one equation from the other, we find

$$\frac{q_1 - q_2}{C} + R(i_1 - i_2) = 0, \tag{3.18}$$

which does not involve induction at all! What happens is that, when the two self-inductions are equal, $M = L$, and the currents flow in opposite directions ($i_1 = -i_2$), the flux created by one solenoid is canceled exactly by the other. There is no flux, no electromotive force produced by induction and the two capacitors discharge into the resistors unimpeded. Students may be asked to find initial conditions that lead to this normal mode.

We now turn to forced oscillations with $k = 1$. When $\nu = 0$ there is only one resonance at $x = 1/\sqrt{2}$, common to $\hat{I}_1$ and $\hat{I}_2$. This is related with the divergence of ω_- as $k \to 1$ discussed above. As ν is increased, one resonant frequency of $\hat{I}_1$ and one resonant frequency of $\hat{I}_2$ come down from very large values of x. Eventually they disappear. The second resonant frequency of $\hat{I}_1$ remains close to $x = 1/\sqrt{2}$. The second resonant frequency of $\hat{I}_2$ coalesces with the first and then grows rapidly.

3.4 Conclusions

We have studied two identical RLC oscillators coupled through mutual induction. Our students predicted correctly that there would be two natural frequencies. But, contrary to our students' intuition: the resonant frequencies do not coincide with the natural frequencies; the resonant frequencies of both currents differ; and, there are even some parameter choices for which only one resonant frequency exists. Our classes found this to be a very helpful exercise.

Acknowledgments

We are very grateful to R. G. Felipe for helping us with some mathematical packages.

References

[1] B. J. Weigman and H. F. Perry, "Experimental determination of normal frequencies in coupled harmonic oscillator system using fast Fourier transforms: An advanced undergraduate laboratory," Am. J. Phys. **61**, 1022–1027 (1993); R. Givens, O. F. Alcantara Bonfima, and R. B. Ormond, "Direct observation of normal modes in coupled oscillators," Am. J. Phys. **71** 87–90 (2003).

[2] W. Frank and P. Brentano, "Classical analogy to quantum mechanical level repulsion," Am. J. Phys. **62** 706–709 (1994).

[3] G. Hansen, O. Harang, and R. J. Armstrong, "Coupled oscillators: A laboratory experiment," Am. J. Phys. **64** 656–660 (1996).

[4] See, for example, I. G. Main, *Vibrations and Waves in Physics*, 3rd ed. (Cambridge University Press, Cambridge, 1994), p. 126; A. P. French, *Vibrations and Waves: The MIT Introductory Physics Series* (W. W. Norton and Co., New York, 1971), p. 135; F. S. Crawford, *Waves: Berkeley Physics Course - vol. 3* (McGraw-Hill Publishing Co., New York, 1968), p. 117.

4

Maximum Thermodynamic Power Coefficient of a Wind Turbine

J. M. Tavares and P. Patrício

CONTENTS

According to the centenary Betz-Joukowsky's law, the power extracted from a wind turbine in open flow can not exceed 16/27 of the wind transported kinetic energy rate. This limit is usually interpreted as an absolute theoretical upper bound for the power coefficient of all wind turbines, but it was derived in the special case of incompressible fluids. Following the same steps of Betz's classical derivation, we model the turbine as an actuator disk in a one dimensional fluid flow, but consider the general case of a compressible reversible fluid, such as air. In doing so, we are obliged to use not only the laws of mechanics, but also and explicitly the laws of thermodynamics. We show that

Originally published as J. M. Tavares and P. Patrício, "Maximum thermodynamic power coefficient of a wind turbine," Wind Energy **23**, 1077–1084 (2020), https://doi.org/10.1002/we.2474, ©2020 John Wiley & Sons, Ltd. Reproduced with permission.

the power coefficient depends on the inlet wind Mach number M_0, and that its maximum value exceeds the Betz-Joukowsky's limit. We have developed a series expansion for the power coefficient in powers of the Mach number M_0 that unifies all the cases (compressible and incompressible) in the same simple expression: $\eta = 16/27 + 8/243 M_0^2$.

4.1 Introduction

The second law of Thermodynamics sets limits to the power coefficient of all energy conversions. Knowledge of the maximum power coefficient of a given conversion process is crucial to increasing its performance. Thermodynamics establishes an hierarchy for the power coefficient η of cyclic heat engines: (i) it is not possible to convert all heat into work, $\eta < 1$; (ii) the power coefficient cannot exceed that of a Carnot cycle between the higher, T_{max}, and lower temperatures, T_{min}, it attains, $\eta < 1 - T_{min}/T_{max}$, independently of other specificities of the engine; (iii) the maximum power coefficient can be calculated for any specific set of cyclic transformations, if these are considered reversible and if the equations of state of the substance that operates the engine are known [1]. Examples of the latter are well known for different cycles of an ideal gas, that emulate several real engines [2]: the Otto cycle (a gasoline engine), the Brayton cycle (a jet engine), the Diesel cycle (a diesel engine), etc.

Nowadays, heat engines are being replaced by renewable energy converters, the most rapidly growing of which are wind turbines for electricity generation [3]. Wind turbines transform the kinetic energy of wind into work with no need for heat sources. The study of their power coefficient dates back to the 1920's when the so called Betz-Joukowsky (BJ) law was derived [4, 5]: by considering the turbine as a thin disc—or actuator—normal to a stationary, non-viscous and incompressible flow, the ratio between the power produced by the turbine and that carried by the fluid flow was shown to have a maximum of 16/27. Although it was derived for incompressible flows, it is not uncommon to find references in the literature to the BJ limit such as "the maximum amount of energy one can get from the wind" [5], "the maximum theoretically possible rotor power coefficient" [6], "theoretical maximum for an ideal wind turbine" [7], "maximum achievable value of the power coefficient" [8]. This suggests that this limit is being interpreted as a kind of Carnot limit: just like any cyclic heat engine has $\eta < 1 - T_{min}/T_{max}$, also any wind turbine would have $\eta < 16/27$, independently of all its specificities.

The derivation of the BJ limit seems to rely only on the laws of mechanics (mass, momentum and energy conservation) [6,9,10]. This is because this result was derived in the context of water propellers and then applied to wind turbines exposed to low velocity air flows [5]: the incompressibility hypothesis, valid in these cases, leads to the decoupling of the mechanical and thermodynamical descriptions of the flow. The purpose of this work is to derive a more general expression for the maximum power coefficient of a wind turbine allowing for the finite compressibility of a fluid, such as air. By doing so, we are obliged to use not only the laws of mechanics, but also and explicitly the laws of thermodynamics. We show that the power coefficient depends on the inlet wind Mach number M_0. While the BJ law is recovered in the limit of zero M_0, it will be shown that the famous limit it sets for power coefficient (16/27) can be surpassed if one considers, e.g., isentropic or isothermal flows.

The actuator disk theory for compressible flows has been previously studied in the 1950's, in the context of propellers [11,12], and more recently, in the context of turbines [13]. Our approach to the problem is very similar to those presented in these studies, and our final results for isentropic flows coincide with the numerical results of [13]. However, in our derivation we explicitly use the second law of thermodynamics and put in evidence both its relation to the calculation of maximum efficiencies and the necessity to consider the entropy in compressible flows. This allows to treat all types of flow (compressible and incompressible) in a unified manner and to consider several types of compressible flows in a systematic way. As a consequence, we analyze not only the isentropic (like [11–13]) but also the isothermal case. Moreover, the final expressions for the efficiency in the compressible flows obtained here are simpler, more clear and more easily related to the Betz limit than those of [11–13]. This simplicity is manifest in the derivation of a novel series expansion for the maximum power coefficient in powers of the Mach number M_0 that unifies all the cases (compressible and incompressible): $\eta = 16/27 + 8/243 M_0^2$. For the isentropic flow, this expression is approximately valid up to $M_0 < 0.8$. In Sec. 4.3.5, we seek for a more detailed explanation to the increase in power coefficient that we did not find in [11–13].

4.2 Power coefficient of a wind turbine

The power coefficient η, or efficiency [4, 9], of a wind turbine is defined as follows [5,6,9,14]. Consider a fluid of mass density ρ_0 in a uniform, horizontal flow with velocity c_0. The kinetic energy per unit time (power) carried by the wind that crosses a vertical planar surface of area A is $P_w = \frac{1}{2}\rho_0 A c_0^3$. A turbine is placed in this surface, distorts the flow, and generates a power $\dot{W}_t$. Its power coefficient is then,

$$\eta = \frac{\dot{W}_t}{P_w}, \tag{4.1}$$

i.e., the fraction of energy that the turbine can take from the flowing air and transform into work. The calculation of $\dot{W}_t$ is done by applying energy conservation to a control volume (CV) that contains the turbine [9]. The first law of thermodynamics applied to CV is,

$$\oint\!\!\oint_{CS}\left(h+\frac{c^2}{2}\right)\rho\vec{c}\cdot\hat{n}dS=\dot{Q}-\dot{W}, \tag{4.2}$$

where CS is the boundary of CV, $\hat{n}$ is the outward-pointing normal to CS, h, ρ and $\vec{c}$ are the enthalpy per unit mass, the mass density and the velocity of the fluid, respectively. $\dot{Q}$ and $\dot{W}$ are the heat and the work per unit time exchanged by the fluid in CV with its vicinity, respectively. The sign convention followed is such that $\dot{Q}<0$ and $\dot{W}>0$ correspond to energy leaving CV to its vicinity. Notice that $h+\frac{c^2}{2}$ is the energy per unit mass transported by a fluid and that $\dot{W}$ represents all the types of work exchanged, i.e., the power delivered by the turbine and the dissipative work of viscous forces. The second law of thermodynamics can be expressed through a generalized Clausius inequality [15] (see supplementary material section) applied to CV,

$$\iiint_{CV} T\rho\vec{c}\cdot\nabla s dV\geq\dot{Q}, \tag{4.3}$$

where ∇s is the gradient of the entropy per unit mass of the fluid and T is the temperature. The equality in Eq. (4.3) is valid for reversible processes, which, in the context of fluid flow, are obtained when viscous forces are absent and the transfer of heat originates by conduction only and associated to a well defined gradient of T. Combination of the first and second laws of thermodynamics, Eqs. (4.2) and (4.3) repectively, results in an expression for the reversible power produced by the turbine,

$$\dot{W}_{t,r}=-\oint\!\!\oint_{CS}\left(h+\frac{c^2}{2}\right)\rho\vec{c}\cdot\hat{n}dS+\iiint_{CV}T\rho\vec{c}\cdot\nabla s dV. \tag{4.4}$$

Since $\dot{W}_{t,r}\geq\dot{W}_t$, this is the expression to be used in the calculation of the maximum power coefficient.

4.3 One-dimensional reversible fluid flows

As in the derivation of the BJ law [5,6,9,14], it is assumed that the distortion of the free flow originated by the placement of the turbine can be represented by a stream tube with cylindrical symmetry surrounding it (see Fig. 7.1). Therefore, all the air that crosses the turbine enters this tube upstream with uniform and horizontal velocity c_0, i.e., the velocity of the free flow, through a planar, vertical section 0 with area A_0 and leaves it downstream through section 3—also planar and vertical with area A_3—with uniform and horizontal velocity c_3. It is also assumed that the thermodynamic state of the air at the inlet (section 0) and the outlet (section 3) is the same, and that the air enters

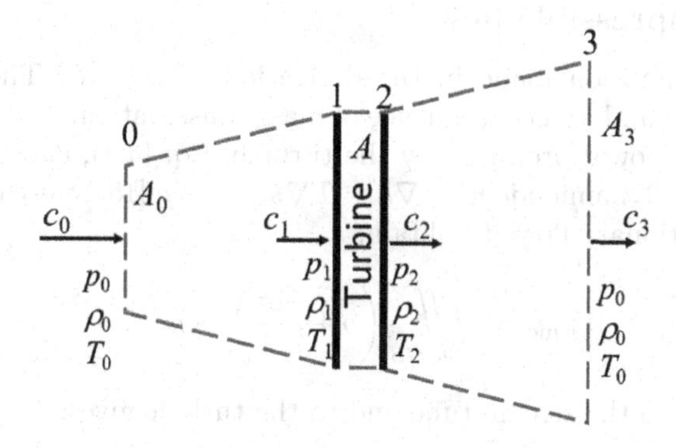

FIGURE 4.1
Schematic representation of the flow around the turbine. A stream tube with cylindrical symmetry around the direction of free flow is defined (region enclosed by the red dashed lines). The air flows into the stream tube by section 0, enters the turbine in section 1, and leaves the turbine and the stream tube at sections 2 and 3, respectively. As a consequence the non vertical red dashed lines are streamlines. At each of these sections, the air is characterized by a velocity c_i whose direction is perpendicular to the plane of the turbine (one dimensional flow approximation) and by the thermodynamic variables p, pressure; ρ, mass density; and T temperature. It is assumed that the thermodynamic states of the air when entering and exiting the stream tube are the same, i.e., $(p_0, \rho_0, T_0) = (p_3, \rho_3, T_3)$.

the turbine with velocity c_1 and leaves it with velocity c_2, both uniform and horizontal. The whole stream tube and some parts of it are going to be considered as control volumes to which the laws of mechanics and thermodynamics can be applied. This will translate into relations between the thermodynamical and mechanical quantities that caractherize the flow, and will allow the calculation of the maximum power coefficient.

Mass conservation implies that the mass flow $\dot{m}$ is the same through all sections of the stream tube and thus,

$$\dot{m} = \rho_0 A_0 c_0 = \rho_1 A c_1 = \rho_2 A c_2 = \rho_3 A_3 c_3. \tag{4.5}$$

The turbine and the stream tube are acted on by an external force that balances the change in linear momentum and pressure of the flowing air. This balance [14] (see supplementary material section) results in,

$$\dot{m}(c_3 - c_0) = (p_2 - p_1)A + \dot{m}(c_2 - c_1). \tag{4.6}$$

The power produced by the turbine, given by Eq. (4.4) and its power coefficient will depend on the type of flow and on the equations of state of the fluid.

4.3.1 Incompressible flow

This is the assumption made in the derivation of BJ law. The density is constant ($\rho_i = \rho$) and as a consequence of mass conservation, $c_1 = c_2 = c$. The equation for the power produced by the turbine, Eq. (4.4), can be simplified using the thermodynamic identity $\nabla h = T\nabla s + \nabla p/\rho$ [15] and the continuity equation for stationary flows to obtain,

$$\dot{W}_{r,\text{inc}} = -\oint\!\!\!\oint_{CS} \left(p + \rho\frac{c^2}{2}\right) \vec{c} \cdot \hat{n} dS. \tag{4.7}$$

The application to the stream tube and to the turbine gives,

$$\dot{W}_{r,\text{inc}} = \dot{m}\frac{c_0^2 - c_3^2}{2}, \tag{4.8}$$

and,

$$\dot{W}_{r,\text{inc}} = (p_1 - p_2)cA. \tag{4.9}$$

The combination of these two equations imposes another relation between the flow properties and A,

$$\dot{m}\frac{c_0^2 - c_3^2}{2} = (p_1 - p_2)cA, \tag{4.10}$$

which combined with the linear momentum equation (4.6) (recall that $c_1 = c_2 = c$) gives,

$$c = \frac{c_0 + c_3}{2}. \tag{4.11}$$

Using $\dot{m} = \rho cA = \rho A(c_0 + c_3)/2$ in Eq. (4.8), the power generated by the turbine is,

$$\dot{W}_{r,\text{inc}} = \rho A\frac{(c_0 + c_3)^2(c_0 - c_3)}{4}. \tag{4.12}$$

The power coefficient of the turbine in an incompressible flow is then,

$$\eta_{r,\text{inc}}(x) = \frac{(1 + x)(1 - x^2)}{2}, \tag{4.13}$$

where $x = c_3/c_0$. This function has its maximal value $\eta_{\text{inc,max}} = 16/27$ when $x = 1/3$. This is BJ law.

4.3.2 Isentropic flow of an ideal gas

Instead of an incompressible fluid, as taken in the derivation of BJ law, let us consider a compressible ideal gas with constant specific heats. As usual, C_p and C_V are the constant pressure and constant volume specific heats, respectively, and $\gamma \equiv C_p/C_V$ is the adiabatic index. The relevant equations of state are $p = \rho rT$ (ideal gas equation), where $r = C_p - C_V$, and $\Delta h = C_p\Delta T$.

For an isentropic flow $\Delta s = 0$ and, for an ideal gas, $T\rho^{1-\gamma} = \text{cte}$. The reversible power produced by the turbine (Eq. (4.4)) becomes:

$$\dot{W}_{r,\text{isen}} = -\oiint_{CS}\left(h + \frac{c^2}{2}\right)\rho\vec{c}\cdot\hat{n}dS. \tag{4.14}$$

The application of Eq. (4.14) to the stream tube and to the turbine gives,

$$\dot{W}_{r,\text{isen}} = \dot{m}\frac{c_0^2 - c_3^2}{2}, \tag{4.15}$$

and,

$$\dot{W}_{r,\text{isen}} = \dot{m}\left(C_p(T_1 - T_2) + \frac{c_1^2 - c_2^2}{2}\right). \tag{4.16}$$

Combining these equations gives,

$$\frac{c_0^2 - c_3^2}{2} = C_p(T_1 - T_2) + \frac{c_1^2 - c_2^2}{2}. \tag{4.17}$$

The linear momentum equation (4.6) can be rewritten using the ideal gas equation and $\dot{m} = \rho_1 c_1 A = \rho_2 c_2 A$, with the result,

$$c_0 - c_3 = \frac{rT_1}{c_1} - \frac{rT_2}{c_2} + c_1 - c_2. \tag{4.18}$$

In order to express Eqs. (4.17, 4.18) as a function of the velocities c_i we use:

(i) a combination of mass conservation in the turbine, $\rho_1 c_1 = \rho_2 c_2$, with the relation between density and temperature in the isentropic flow along the turbine, $T_1\rho_1^{1-\gamma} = T_2\rho_2^{1-\gamma}$, which holds,

$$\frac{T_2}{T_1} = \left(\frac{c_2}{c_1}\right)^{1-\gamma}. \tag{4.19}$$

(ii) the definition of the Mach number M, for the free flow,

$$T_0 = \frac{c_0^2}{r\gamma M_0^2}. \tag{4.20}$$

(iii) conservation of energy applied to the control volume defined by sections 0 and 1,

$$T_1 = T_0 + \frac{c_0^2 - c_1^2}{2C_p} = \frac{c_0^2}{r\gamma M_0^2} + \frac{c_0^2 - c_1^2}{2C_p}. \tag{4.21}$$

Using Eqs. (4.19) and (4.21) to eliminate T_1 and T_2 from Eqs. (4.17, 4.18), these equations become a relation between the velocities c_i and the Mach number of the free flow, M_0:

$$1 - x^2 = \left(\frac{2}{(\gamma-1)M_0^2} + 1 - y^2\right)(1 - z^{1-\gamma}) + y^2(1 - z^2), \tag{4.22}$$

$$(1-x)y = \left(\frac{1}{\gamma M_0^2} + \frac{\gamma-1}{2\gamma}(1-y^2)\right)(1-z^{-\gamma}) + y^2(1-z), \tag{4.23}$$

where $x = c_3/c_0$, $y = c_1/c_0$ and $z = c_2/c_1$. These equations allow the calculation of y and z as a function of x for a given M_0. The power coefficient of the turbine may be written, using Eq. (4.15) and $\dot{m} = \rho_1 c_1 A$, as,

$$\eta_{r,\text{isen}} = \frac{\rho_1 c_1}{\rho_0 c_0}\left(1 - \frac{c_3^2}{c_0^2}\right). \tag{4.24}$$

Using the relation for the isentropic flow $\rho_1/\rho_0 = (T_1/T_0)^{\frac{1}{\gamma-1}}$, Eq. (4.21) and the definitions of x and y, the power coefficient of the turbine in an isentropic flow of an ideal gas is obtained as a function of x and M_0 (for a given γ),

$$\eta_{r,\text{isen}}(x, M_0) = \left(1 + \frac{\gamma-1}{2}M_0^2(1-y^2)\right)^{\frac{1}{\gamma-1}} y(1-x^2). \tag{4.25}$$

Notice that y is a function of x and M_0 given by Eqs. (4.22) and (4.23), that may be computed numerically for the general case.

However, a perturbative analysis gives us also interesting results. To first order in M_0^2, we have

$$y(x, M_0^2) = \frac{1+x}{2} - \frac{M_0^2}{8}(1-x)(1+x)^2 + ... \tag{4.26}$$

The zero-th order term leads to the Betz power coefficient (see Eq. (4.13)), whose maximum value $\eta_{\text{inc,max}} = 16/27$ is attained when $x = 1/3$. To first order in M_0^2, the maximum power coefficient (also for $x = 1/3$) is given by

$$\eta_{r,\text{isen,max}} = \frac{16}{27} + \frac{8}{243}M_0^2 + ... \tag{4.27}$$

We notice that this result does not depend on γ, but only on M_0.

4.3.3 Isothermal flow of an ideal gas

For an isothermal flow (T = cte) of an ideal gas, the reversible power produced by the turbine (Eq. (4.4)) is,

$$\dot{W}_{r,\text{isot}} = -\oint\!\!\oint_{CS}\left(\frac{c^2}{2} - Ts\right)\rho\vec{c}\cdot\hat{n}dS. \tag{4.28}$$

Applying this equation to the stream tube and the turbine gives,

$$\dot{W}_{r,\text{isot}} = \dot{m}\frac{c_0^2 - c_3^2}{2}. \tag{4.29}$$

$$\dot{W}_{r,\text{isot}} = \dot{m}T(s_2 - s_1) + \dot{m}\frac{c_1^2 - c_2^2}{2}. \tag{4.30}$$

In an ideal gas with constant specific heats, the specific entropy variation in an isothermal process is $s_2 - s_1 = r \ln \frac{p_1}{p_2} = r \ln \frac{\rho_1}{\rho_2}$. Conservation of mass in the turbine reads $\rho_1 c_1 = \rho_2 c_2$, leading to,

$$\dot{W}_{isot} = \dot{m} r T \ln \frac{c_2}{c_1} + \dot{m} \frac{c_1^2 - c_2^2}{2}, \tag{4.31}$$

which, combined with Eq. (4.29) gives,

$$\frac{c_0^2 - c_3^2}{2} = rT \ln \frac{c_2}{c_1} + \frac{c_1^2 - c_2^2}{2}. \tag{4.32}$$

After considering that $\dot{m} = \rho_1 c_1 A = \rho_2 c_2 A$ and $p_i/\rho_i = rT$, the linear momentum equation (4.6) for an isothermal flow can be expressed as,

$$c_0 - c_3 = rT \left(\frac{1}{c_1} - \frac{1}{c_2} \right) + c_1 - c_2. \tag{4.33}$$

Equations (4.32) and (4.33) can be expressed as,

$$(1 - x^2) \gamma M_0^2 = 2 \ln z + \gamma M_0^2 y^2 (1 - z^2), \tag{4.34}$$

$$\gamma M_0^2 y z \left(1 - x - y(1 - z)\right) = z - 1, \tag{4.35}$$

where, as before, $x = c_3/c_0$, $y = c_1/c_0$, $z = c_2/c_1$ and $rT = c_0^2/\gamma/M_0^2$. These two equations allow the calculation of y, z for fixed x, M_0.

The power coefficient for an isothermal flow is then,

$$\eta_{r,\text{isot}} = \frac{\rho_1 A c_1 (c_0^2 - c_3^2)}{\rho_0 A c_0^3} = \frac{\rho_1}{\rho_0} y (1 - x^2). \tag{4.36}$$

In order to determine ρ_1/ρ_0 as a function of the velocities c_i, we use the conservation of energy for the flow between sections 0 and 1 of the stream tube to obtain

$$\frac{c_1^2 - c_0^2}{2} = T(s_1 - s_0) = -rT \ln \frac{\rho_1}{\rho_0}, \tag{4.37}$$

from which,

$$\frac{\rho_1}{\rho_0} = \exp \left(\frac{c_0^2 - c_1^2}{2rT} \right). \tag{4.38}$$

The power coefficient can then be expressed as a function of x and M_0,

$$\eta_{r,\text{isot}}(x, M_0) = \exp \left(\frac{\gamma M_0^2}{2} (1 - y^2) \right) y (1 - x^2), \tag{4.39}$$

with $y(x, M_0)$ given by Eqs. (4.34) and (4.35). The perturbative analysis leads to the same maximum power coefficient as is the isentropic case, to first order in M_0^2 (it had been shown already that it did not depend on γ).

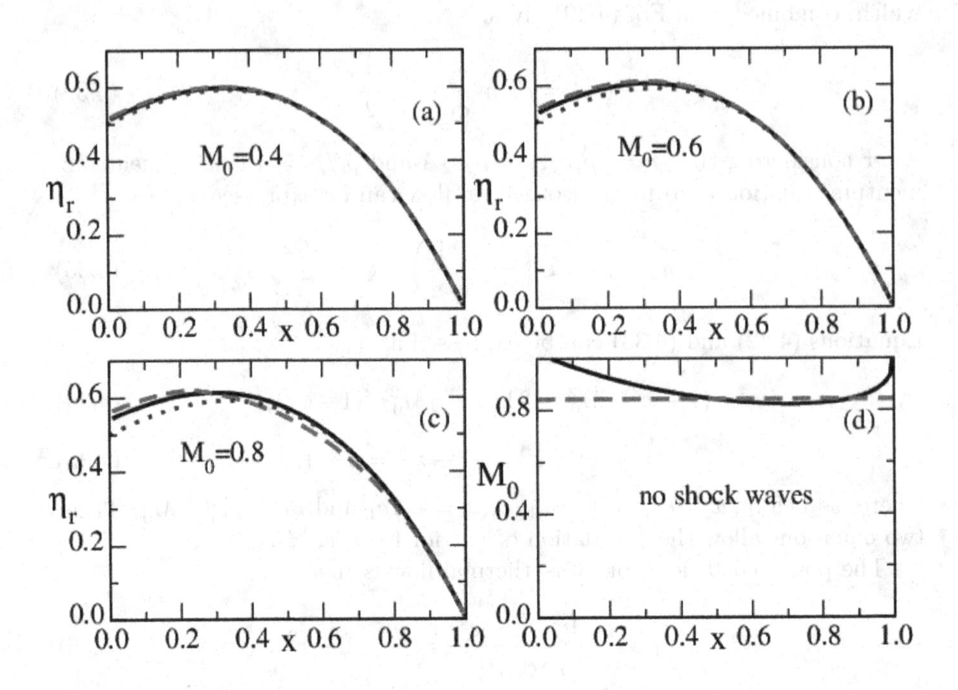

FIGURE 4.2
(a) to (c): reversible power coefficient η_r as a function of $x = c_3/c_0$ for incompressible (blue dotted line), isentropic (black full line), and isothermal (red dashed line) flow, for the indicated values of M_0, the Mach number of the free flow; (d): for the pairs of values (x, M_0) above the black full (red dashed) line, isentropic (isothermal reversible) flow is no longer possible due to the formation of shock waves. The isentropic and isothermal flows are that of an ideal gas with adiabatic index $\gamma = 1.4$.

4.3.4 Power coefficient calculations

The power produced by the turbine (4.4) and its power coefficient will depend on the type of flow and on the equations of state of the fluid. We have combined Eqs. (4.4),(4.5) and (4.6) in three cases: (i) incompressible flow ($\rho = \text{cte}$), (ii) isentropic flow ($s = \text{cte}$), and (iii) isothermal flow ($T = \text{cte}$) of an ideal gas. In case (i), the BJ law is recovered: the power coefficient is a function of the ratio $x = c_3/c_0$ only, and is given by a single expression, $\eta_{r,\text{inc}} = (1+x)(1-x^2)/2$ (Eq. (4.13)). In cases (ii) and (iii), we obtain $\eta_{r,\text{isen}}(x, M_0, \gamma)$ and $\eta_{r,\text{isot}}(x, M_0, \gamma)$: the efficiencies depend not only on x, but also on the Mach number of the free flow, M_0, and on the adiabatic index of the gas γ. To calculate them, we must solve a system of three Eqs. (4.22), (4.23) and (4.25) in case (ii) and (4.34), (4.35) and (4.39) in case (iii)).

The efficiencies for these three flows are computed as a function of x and shown in Fig. 4.2(a)–(c) for 3 values of M_0 (and for $\gamma = 1.4$). For low values of M_0 (up to ≈ 0.5), the efficiencies are almost indistinguishable: this means that the density variations are negligible, and the incompressible approximation holds. For higher values of M_0, the efficiencies become noticeable different: the flow is in the subsonic regime, where density variations are important but shock waves are not present. For values of M_0 close to the supersonic regime (larger than ≈ 0.8), it is not possible to compute η_r for all values of x, since the assumption of reversibility becomes invalid when shock waves appear in some part of the flow. For each type of compressible flow, the plane (x, M_0) is divided (for a given γ) into two regions (see Fig. 4.2(d)): only in the no shock wave region is it possible to define η_r. In this region of parameters all M_i (Mach number of the flow in section i of the CV—see Fig. 4.1) are < 1 for isentropic flows and $< 1/\sqrt{\gamma}$ for isothermal flows. The lines in Fig. 4.2(d) were calculated: (i) for isentropic flows by solving $M_2 = 1$, since M_2 is always the largest Mach number; (ii) for isothermal flows by setting $M_0 = 1/\sqrt{\gamma}$, since $M_i < 1/\sqrt{\gamma}$ when $M_0 < 1/\sqrt{\gamma}$.

The maximum power coefficient of turbines in each of the three cases is obtained by determining the maximum of the functions $\eta_r(x)$. In the incompressible case, this calculation is done analytically and the BJ limit is recovered: $\eta_{\text{inc,max}} = 16/27$. The results of the numerical calculation for the maximum power coefficient of isentropic ($\eta_{\text{isen,max}}(M_0)$) and isothermal ($\eta_{\text{isot,max}}(M_0)$) turbines for each value of M_0 are plotted in Fig. 4.3 and compared to $\eta_{\text{inc,max}}$. The maximum efficiencies in these cases are larger than 16/27 at all values of M_0 in the isothermal case and for M_0 up to ≈ 0.95 in the isentropic case. At high Mach numbers of the free flow ($M_0 \approx 0.9$ for isentropic flows and $M_0 \approx 0.8$ for isothermal flows) the power coefficient is increased by 4–5%. The perturbative expansion to first order in M_0^2, $\eta_{r,\text{isen,max}} \approx \eta_{r,\text{isot,max}} \approx 16/27 + 8/243 M_0^2$ is also shown. Notice that this quadratic expression, valid for small values of M_0, follows approximately the maximum efficiency for the isentropic case up to $M_0 < 0.8$.

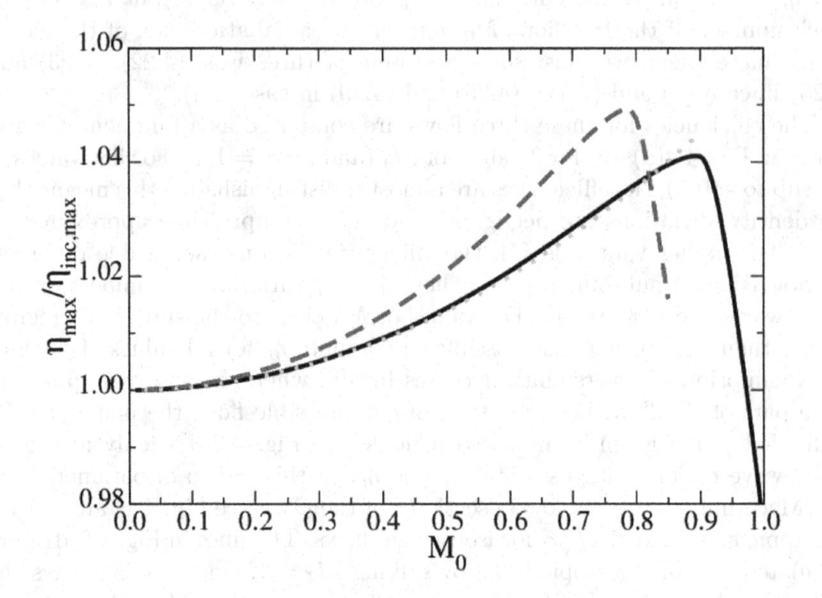

FIGURE 4.3
Maximum power coefficient of isentropic (full black line) and isothermal (dashed red line) turbines compared to the Betz limit (incompressible case, $\eta_{\text{inc,max}} = 16/27$) as a function of the Mach number of free flow, M_0. The green dotted line represents the Taylor expansion of $\eta_{\text{max}}/\eta_{\text{inc,max}}$ to first order in M_0^2 (see the text).

These results suggest a reinterpretation of the BJ limit as a convenient, practical and easily derivable value for maximum wind turbine power coefficient, that is always surpassed when compressibility is taken into account.

4.3.5 Analysis

The origin of the increase in power coefficient due to compressibility can be better understood when the power coefficient is considered as a product of two factors, $\eta_r = \alpha \times \beta$. α is the ratio between the mass flow that crosses the turbine and the mass flow of the wind,

$$\alpha = \frac{\rho_1 c_1}{\rho_0 c_0}. \tag{4.40}$$

β is the fraction of energy per unit mass of fluid that the turbine extracts from the free flow,

$$\beta = 1 - \frac{c_3^2}{c_0^2} = 1 - x^2 \tag{4.41}$$

The maximum power coefficient in incompressible flow is obtained when $\alpha_{\text{inc}} = 2/3$ and $\beta_{\text{inc}} = 8/9$: the mass flow arriving to the turbine is 2/3 of that of free flow; the turbine then converts into work 8/9 of the kinetic energy contained in each unit mass of the fluid. Figure 4.4 shows the results for α and β in compressible flows, in conditions of maximum power coefficient as a function of M_0. The behavior is similar in both isentropic and isothermal flows: β grows continuously with increasing M_0 while α has a slower increase and a sudden decrease at high M_0. Except in the case of α at very large M_0, both coefficients are larger than those of incompressible flow. Therefore, the power coefficient of compressible flows is larger than that of the incompressible limit, because compressibility leads to a larger amount of fluid arriving to the turbine ($\alpha > 2/3$) and, specially, to the ability of the turbine to extract more work from the flow ($\beta > 8/9$). Notice that an increase over the Betz limit in power coefficient can also be obtained in incompressible flow by surrounding the turbine with a diffuser [16] (the so called diffuser augmented wind turbines—DAWT). However, in DAWT's this increase is solely due to an increase in α [16]. On the contrary, the largest contribution to the increase of the maximum power coefficient in compressible flow comes from the increase in β, i.e., from the increase in the efficiency of the turbine when converting the actual wind kinetic energy that reaches it into work (see Fig. 4.4).

4.4 Conclusion

The analysis presented in this work clarifies the status of the BJ limit: it is a theoretical upper bound for the power coefficient of reversible wind

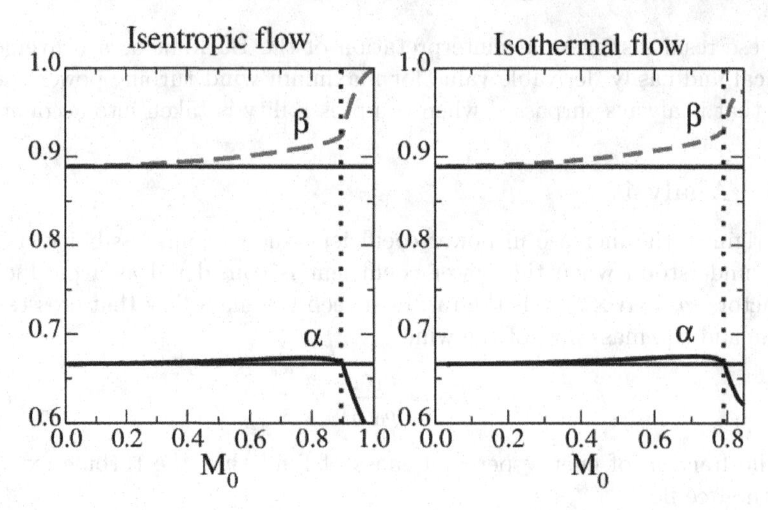

FIGURE 4.4
Components α, given by Eq. (4.40), and β, given by Eq. (4.41), of the maximum power coefficient of wind turbines. The horizontal full lines are the values $\alpha = 2/3$ and $\beta = 8/9$ for the incompressible case. The dotted vertical lines mark the value of M_0 for which the power coefficient (the product $\alpha \times \beta$) is maximum.

turbines within the incompressibility hypothesis. Once this hypothesis is dropped, larger efficiencies can be attained. The upper limit will depend on the type of flow and on the equation of state of the fluid. Our results also strongly suggest that the BJ limit is in fact the smallest maximum possible power coefficient, since it appears to be reached from above in the limit $M_0 \to 0$ of maximum efficiencies of compressible flows. Even if these results are expected to have small practical value for the real systems in current use, they clarify the role of general scientific principles (like the first and second law of thermodynamics) in the energy conversion occurring in wind turbines. The new results of this paper are condensed in the simple form of a series expansion of the maximum power coefficient to first order in M_0^2, $\eta_{max} = 16/27 + 8/243M_0^2$, valid for M_0 up to ≈ 0.8 and for ideal isentropic gas flows. This simple equation contains the BJ limit (recovered when $M_0 = 0$) and shows that compressibility increases the efficiency but only a few percent when compared to the incompressible case.

Notice that the role of compressibility has been neglected in the study of real (irreversible) wind turbines, since they operate at low air velocities (e.g., they are shut down when wind velocity exceeds ≈ 20 m/s [6] or $M_0 \approx 0.06$). However, at the tip of the large and more modern wind turbines blades air can reach velocities that correspond to $M_0 \approx 0.2-0.3$ and so the effects of compressibility on the wake of the turbine and on its power production are starting to be explored through more realistic models and more sophisticated numerical methods [17]. Although one should not forget that for higher Mach numbers, turbulence effects increase the irreversibility of the flow, it would be important to extend the present analysis to a model that considers the rotation of the turbine and its size, i.e., the actuator disk is replaced by a rotor that creates a wake in the air flow [6, 14, 18]. The power coefficient of reversible turbines with a given tip speed ratio—the ratio of the velocity of the tip of the turbine to the wind velocity—in compressible flows should be a reference for more realistic irreversible flows. It could also be compared with the known results for the incompressible case.

4.5 Supplementary material

4.5.1 Generalized clausius inequality

Let us consider an elementary volume dV of a control volume CV, whose temperature is T and density is ρ. Suppose that the vector heat transfer per unit area [9] in this volume is $\vec{q}$, and the entropy is S. Then the second law of thermodynamics relates T, $\vec{q}$ and the variation of S with time,

$$T\frac{dS}{dt} \geq -\nabla \cdot \vec{q}dV. \tag{4.42}$$

Using $S = \rho s dV$ and the continuity equation, this becomes,

$$T\rho\frac{ds}{dt}dV \geq -\nabla \cdot \vec{q}dV, \tag{4.43}$$

which, integrated over the whole CV, gives

$$\iiint_{CV} T\rho\frac{ds}{dt}dV \geq \dot{Q}, \tag{4.44}$$

since

$$\dot{Q} \equiv -\iiint_{CV} \nabla \cdot \vec{q}dV = -\oiint_{CS} \vec{q} \cdot \hat{n}dA. \tag{4.45}$$

For a stationary flow, Eq. (4.44) can be rewritten as,

$$\iiint_{CV} T\rho\vec{c} \cdot \nabla s dV \geq \dot{Q}. \tag{4.46}$$

The entropy variation with time can be related to the heat flow and to the dissipation through viscosity [15]. The equality in Eq. (4.46) is obtained when heat conduction is due to a defined gradient of temperature and the fluid viscosity is zero.

4.5.2 Linear momentum equation

The linear momentum equation for a fixed control volume CV, and an inviscid fluid is,

$$\vec{F}_{ext} - \oiint_{CS} p\hat{n}dS = \oiint_{CS} \vec{c}\rho(\vec{c}\cdot\hat{n})dS. \tag{4.47}$$

where CS is the control surface of CV and $\vec{F}_{ext}$ is the external force acting on it. The momentum equation is applied to the stream tube in the direction z of the free flow to obtain,

$$F_{ext,z} = \oiint_{CS} p\hat{n}_z dS + \dot{m}(c_3 - c_0), \tag{4.48}$$

where $F_{ext,z}$ and $\hat{n}_z$ are the z component of the external force acting on the stream tube and of the outer normal to its control surface. Glauert has shown that $\oiint_{CS} p\hat{n}_z dS = 0$ for an incompressible inviscid fluid [14]. We show below that this equality also holds for compressible ideal fluids, and thus,

$$F_{ext,z} = \dot{m}(c_3 - c_0). \tag{4.49}$$

The momentum equation can also be applied to the turbine. Since the external force acting on the whole control volume is the same force that acts on the turbine,

$$F_{ext,z} = (p_2 - p_1)A + \dot{m}(c_2 - c_1). \tag{4.50}$$

Combining Eqs. (4.49) and (4.50) the linear momentum equation imposes the following relation between the quantities pertaining to the flow and the area A of the turbine,

$$\dot{m}(c_3 - c_0) = (p_2 - p_1)A + \dot{m}(c_2 - c_1). \tag{4.51}$$

4.5.3 Proof that $\oiint_{CV} p\hat{n}_z dS = 0$ for a compressible ideal flow

Glauert [14] has already proved that $\oiint_{CV} p\hat{n}_z dS = 0$ for an incompressible inviscid fluid. In this section, we will follow his analysis and extend it to the case of a compressible ideal gas with constant specific heats. As usual, C_p and C_V are the constant pressure and constant volume specific heats, respectively, and $\gamma \equiv C_p/C_V$ is the adiabatic index. The relevant equations of state are $p = \rho r T$, where $r = C_p - C_V$, and $\Delta h = C_p \Delta T$. For an isentropic flow $\Delta s = 0$ and, for an ideal gas, $T\rho^{1-\gamma} = \text{cte}$; for an isothermal flow $\Delta T = 0$ and, for an ideal gas, $p/\rho = \text{cte}$.

We represent the flow as before (see Fig. 4.5), but instead of having an infinite surrounding atmosphere, we require that the ideal gas flows in a large horizontal (perpendicular to the inlet velocity) container of section S, and we will take $S \to \infty$. Although the velocity of the fluid varies continuously over the section, we will have $c_3 < c_0$ for the representative average value at

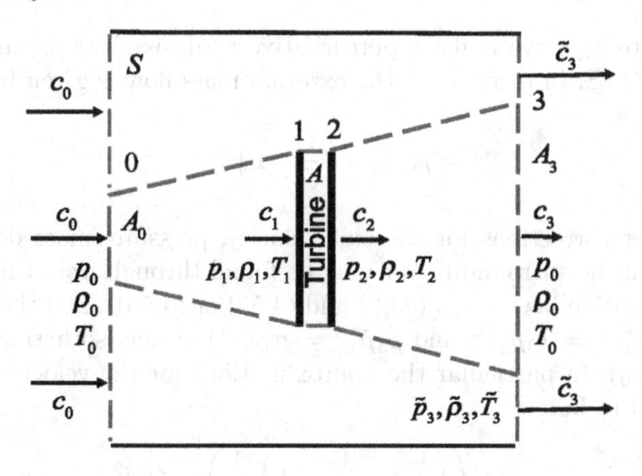

FIGURE 4.5
Schematic representation of the flow around the turbine, but enclosed in a large container of section S. At exit, the surrounding air that crosses section $(S - A_3)$ has an average velocity $\tilde{c}_3$, pressure $\tilde{p}_3$, mass density $\tilde{\rho}_3$, and temperature $\tilde{T}_3$.

section A_3. This means that outside this section, at $(S-A_3)$, the representative average velocity of the fluid is $\tilde{c}_3 > c_0$, in order to ensure mass conservation. However, this velocity $\tilde{c}_3 \to c_0$ as S becomes larger and larger.

Conservation of mass requires that the mass flow inside and outside the turbine are given respectively by $\dot{m} = \rho_0 A_0 c_0 = \rho_3 A_3 c_3$ and

$$\dot{M} = \rho_0 c_0 (S - A_0) = \tilde{\rho}_3 \tilde{c}_3 (S - A_3). \tag{4.52}$$

The rate of change of the linear momentum of the air that crosses the trubine is equal to the sum of all applied forces:

$$\dot{m}(c_3 - c_0) = F_{ext,z} + F_{s,z} + p_0(A_0 - A_3) \tag{4.53}$$

where $F_{ext,z}$ is the force exerted by the turbine, and $F_{s,z}$ is the force exerted by the surroundings (both in the direction of the flow). The latter may be estimated by calculating the rate of change of linear momentum of the flow that surrounds the turbine:

$$\dot{M}(\tilde{c}_3 - c_0) = -F_{s,z} + p_0(S - A_0) - \tilde{p}_3(S - A_3) \tag{4.54}$$

The energy variation in the surrounding fluid is, for adiabatic flows,

$$\frac{1}{2}(\tilde{c}_3^2 - c_0^2) + C_p(\tilde{T}_3 - T_0) = 0, \tag{4.55}$$

and for isothermal and reversible flows,

$$\frac{1}{2}(\tilde{c}_3^2 - c_0^2) = T(\tilde{s}_3 - s_0), \tag{4.56}$$

To estimate $F_{s,z}$, we make a perturbative analysis around the small parameter $\delta = A_0/S$. In terms of δ, the external mass flow is given by

$$\dot{M} = \rho_0 c_0 A_0 \left(\frac{1}{\delta} - 1\right) \tag{4.57}$$

The first order corrections for the exit velocity, pressure, mass density, and temperature of the surrounding air may be found through from the mass and energy conservation laws (Eqs. (4.52) and (4.55) or (4.56)), and the isentropic relations ($\tilde{T}_3\tilde{\rho}_3^{1-\gamma} = T_0\rho_0^{1-\gamma}$ and $\tilde{p}_3\tilde{\rho}_3^{-\gamma} = p_0\rho_0^{-\gamma}$) or the isothermal relations ($\tilde{p}_3/p_0 = \tilde{\rho}_3/\rho_0$). In particular the approximations for the velocity and pressure are found to be

$$\tilde{c}_3 = c_0\left(1 + \delta\left(\frac{A_3}{A_0} - 1\right)\frac{1}{B}\right) + O(\delta^2) \tag{4.58}$$

$$\tilde{p}_3 = p_0\left(1 - \delta\gamma\left(\frac{A_3}{A_0} - 1\right)\frac{M_0^2}{B}\right) + O(\delta^2) \tag{4.59}$$

where $M_0 = c_0/\sqrt{r\gamma T_0}$ is the Mach number and $B = 1 - M_0^2$ (isentropic case) or $B = 1 - \gamma M_0^2$ (isothermal case). Replacing these expressions in Eq. (4.54), we obtain, after several simplifications, for both isentropic and isothermal flows,

$$F_{s,z} = p_0(A_3 - A_0) + O(\delta). \tag{4.60}$$

In the limit $\delta \to 0$, this finally leads to

$$-\oiint_{CV} p\hat{n}_z dS = F_{s,z} + p_0(A_0 - A_3) = 0 \tag{4.61}$$

Acknowledgments

We thank P. I. C. Teixeira and N. A. M. Araújo for a critical reading of this manuscript.

References

[1] H. B. Callen, *Thermodynamics and an Introduction to Thermostatistics*, 2nd ed. (John Wiley and Sons, New York, 1985).

[2] Y. A. Çengel and M. A. Boles, *Thermodynamics: An Engineering Approach*, 8th ed. (McGraw Hill, New York, 2015).

[3] Global Wind Report. Global Wind Energy Council. https://gwec.net/; 2017.

[4] V. L. Okulov and G. A. van Kuik. "The Betz–Joukowsky limit: on the contribution to rotor aerodynamics by the British, German and Russian scientific schools," Wind Energ. **15**, 335–344 (2012).

[5] A. Betz, "The maximum of the theoretically possible exploitation of wind by means of a wind motor," Wind Eng. **37**, 441–446 (2013). Translation of Zeitschrift für das gesamte Turbinenwesen **26**, 307 (1920), by H. Hamann, J. Thayer and A. P. Schaffarczyk.

[6] J. F. Manwell, J. G. McGowan, and A. L. Rogers, *Wind Energy Explained: Theory, Design and Application* (John Wiley and Sons, West Sussex, 2009).

[7] M. O. L. Hansen, *Aerodynamics of Wind Turbines*, 2nd ed. (Earthscan, West Sussex, 2008).

[8] T. Burton, D. Sharpe, N. Jenkins, and E. Bossanyi, *Wind Energy. Handbook* (John Wiley and Sons, London, 2001).

[9] F. M. White. *Fluid Mechanics*, 7th ed. (McGraw-Hill, New York, 2011).

[10] M. Huleihila, "A comparative analysis of deep level emission in ZnO layers deposited by various methods," J.Appl. Phys. **105**, 104908 (2009).

[11] A. W. Vogeley, *Axial-momentum theory for propellers in compressible flow*, NACA Technical Note 2164 (National Advisory Committee for Aeronautics, Langley Aeronautical Lab., Langley Field, VA, 1951).

[12] J. B. Delano and J. L. Crigler, NACA RM L53A07 (1953).

[13] H. H. N. Oo, Actuator disk theory for compressible flow, MSc dissertation, California Polytechnic State University (USA), 2017.

[14] H. Glauert, "Airplane Propellers," in *Aerodynamic Theory*, edited by W. F. Durand (Springer Verlag, Berlin, 1935), pp. 169–360.

[15] L. D. Landau and E. M. Lifshitz, *Fluid Mechanics*, 2nd ed. (Pergamon Press, Oxford, 1987).

[16] S. Hjort and H. Larsen, "A multi-element diffuser augmented wind turbine," Energies **7**, 3256–3281 (2014).

[17] C. Yan and C. L. Archer, "Assessing compressibility effects on the performance of large horizontal-axis wind turbines," Applied Energy **212**, 33–45 (2018).

[18] D. F. Hunsaker and W. F. Philips, "Momentum theory with slipstream rotation applied to wind turbines," in 31st AIAA Applied Aerodynamics Conference 2013, San Diego, California: AIAA–2013–3161.

Part II

Meditations

5

Mutual Inductance between Piecewise Linear Loops

A. C. Barroso and J. P. Silva

CONTENTS

We consider a current-carrying wire loop made out of linear segments of arbitrary sizes and directions in three-dimensional space. We develop expressions to calculate its vector potential and magnetic field in all points in space. We then calculate the mutual inductance between two such (non-intersecting) piecewise linear loops. As simple applications, we consider in detail the mutual inductance between two square wires of equal length which: i) lie on the same plane; ii) lie on parallel horizontal planes, with their centers lying on the same vertical axis. Our expressions can also be used to obtain approximations to the mutual inductance between wires of arbitrary three-dimensional shapes.

Reproduced from A. C. Barroso and J. P. Silva, "Mutual inductance between piecewise linear loops," American Journal of Physics **81**, 829–835 (2013), https://doi.org/10.1119/1.4818278, with the permission of the American Association of Physics Teachers.

5.1 Introduction

All courses on electromagnetism (introductory [1], intermediate [2], or advanced [3]) contain exercises involving Biot-Savart's law. Once the magnetic field is known, whether through Biot-Savart's or through Ampere's law, one can then determine self and mutual inductances; a subject usually addressed in the context of Faraday's law. Most often one finds very simple geometries: the determination of the magnetic field generated by an infinite straight wire, by a finite straight wire (at some particular points), or the field of a circular loop at points along its axis. The literature has been enhanced by other interesting examples of planar wires, such as calculations of the field generated by: an arbitrarily shaped planar wire, in the plane of the wire [4]; or a regular planar polygon, at any point in space [5]. The field of a planar circular loop at any point in space has been studied numerically [6], experimentally [7], and can be deduced from the expression for the vector potential **A** found in the classical textbook by Jackson [3].

The calculations of mutual inductances are also usually presented in the simplest of cases. For example, two coaxial infinite solenoids, two coaxial toroids of rectangular section, or the infinite coaxial cable. A calculation of the mutual inductance between two concentric regular polygons lying on the same plane has also been performed [8].

In this article, we consider two current-carrying wire loops made out of linear segments of arbitrary directions and sizes in three dimensions. One example is presented in Fig. 5.1. We start by calculating the magnetic vector

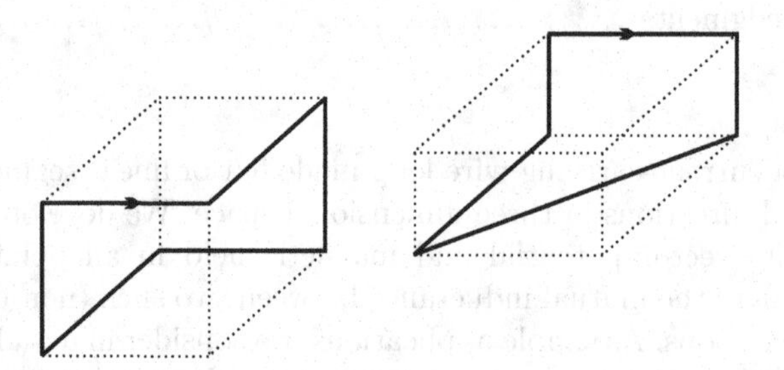

FIGURE 5.1
Two current-carrying piecewise linear loops. The dotted lines represent solid objects which help visualize the structure of the wires in three-dimensional space.

field **A** due to one loop, from which one can deduce its magnetic field $\mathbf{B} = \nabla \times \mathbf{A}$. Our final objective is to determine the mutual inductance between two current-carrying wire loops of arbitrary piecewise linear shapes, such as the ones in Fig. 5.1.

We calculate the flux of the magnetic field $\mathbf{B}$ through a surface S with edges on the curve C described by the second wire loop, as

$$\Phi = \int_S \mathbf{B} \cdot d\mathbf{a} = \int_S (\nabla \times \mathbf{A}) \cdot d\mathbf{a} = \oint_C \mathbf{A} \cdot dl. \tag{5.1}$$

The last equality has the advantage of requiring only a one dimensional integration. Many authors barely mention the vector potential; but, as this problem illustrates, the vector potential can be very useful.

Our work generalizes previous results in the following ways: the wires need not be regular polygons or circular; if planar, the two loops need not be concentric and they may not even lie on the same plane; the wires need not be planar and can describe any piecewise linear figure in three dimensions. Since any sufficiently smooth curve in three-dimensional space may be approximated by a piecewise linear curve, our expressions can also be applied to obtain approximate results for the mutual inductance of any two wires of arbitrary shapes. The work can be applied in the classroom to develop intuition about the dependence of mutual inductances on relative position and/or relative orientation. Advanced students may work with the formulas directly; others may have access to a plotting program where the formulas are already coded. One can also build experiments to test Faraday's law in a huge variety of situations. The results may also be of technological interest. For example, Santra *et.al.* [9] have used the results in Refs. [5] and [8] in order to find the magnetic field created by planar micro-electromagnet spirals for microelectromechanical systems applications (usually known as MEMS). They approximated the spirals by a collection of squares. That approximation can be removed, and we can now calculate the mutual inductance between two such circuits.

In Sec. 5.2, we calculate the vector potential that a finite straight wire generates at all points in space. Such finite wires can be combined to obtain the expression for the vector potential of any piecewise linear wire and, thus, for its magnetic field $\mathbf{B}$. We calculate in Sec. 5.3.1 the line integral of $\mathbf{A}$ along some generic finite straight path, specifying in Sec. 5.3.2 to the simplifying particular case of planar wires. The results obtained are used in Sec. 5.4 to determine the flux that crosses some piecewise linear three-dimensional wire, due to the field created by another, non-intersecting, piecewise linear three-dimensional wire. In Secs. 5.5 and 5.6, we use our results to calculate the mutual inductance of two square loops lying on the same plane and on parallel planes (with the same axis), respectively. We draw our conclusions in Sec. 5.7 and relegate some longer expressions to the appendix.

5.2 The vector potential

Figure 5.2 shows a straight finite circuit, carrying the current i, having length s_k, centered at

$$\mathbf{R}_k = X_{0k}\,\hat{\mathbf{e}}_x + Y_{0k}\,\hat{\mathbf{e}}_y + Z_{0k}\,\hat{\mathbf{e}}_z, \tag{5.2}$$

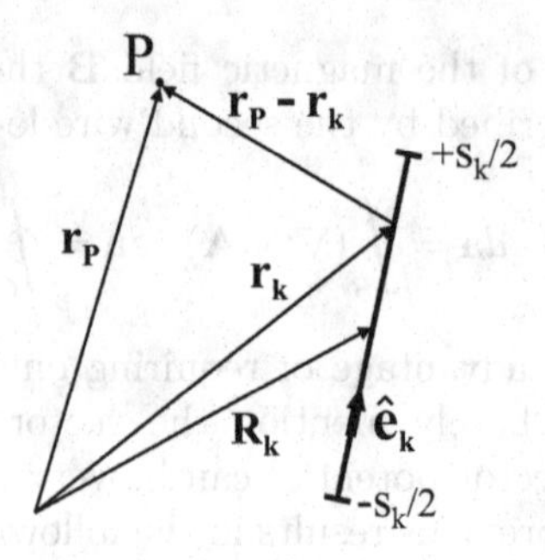

FIGURE 5.2
The finite straight circuit on the right carries a current i. Also shown is the point P where we wish to calculate the vector potential.

and having the direction of the unit vector

$$\hat{\mathbf{e}}_k = a_x \hat{\mathbf{e}}_x + a_y \hat{\mathbf{e}}_y + a_z \hat{\mathbf{e}}_z. \tag{5.3}$$

The points along the wire may be parametrized as

$$\begin{aligned} \mathbf{r}_k(t) &= \mathbf{R}_k + t\,\hat{\mathbf{e}}_k & (5.4)\\ &= [X_{0k} + t\,a_x]\,\hat{\mathbf{e}}_x + [Y_{0k} + t\,a_y]\,\hat{\mathbf{e}}_y + [Z_{0k} + t\,a_z]\,\hat{\mathbf{e}}_z & (5.5)\\ &\equiv X_k(t)\,\hat{\mathbf{e}}_x + Y_k(t)\,\hat{\mathbf{e}}_y + Z_k(t)\,\hat{\mathbf{e}}_z, & (5.6) \end{aligned}$$

where $t \in (-s_k/2, s_k/2)$. Notice that

$$\frac{d\mathbf{r}_k}{dt} = \hat{\mathbf{e}}_k \tag{5.7}$$

is a unit vector in the direction of the current in the straight wire, as required. We wish to calculate $\mathbf{A}_k(\mathbf{r}_P)$ at the point P defined by the vector $\mathbf{r}_P = x\,\hat{\mathbf{e}}_x + y\,\hat{\mathbf{e}}_y + z\,\hat{\mathbf{e}}_z$. Since the distance from each elementary source to the point P is $\mathbf{r}_P - \mathbf{r}_k$, we find

$$\begin{aligned} \frac{4\pi}{\mu_0 i}\mathbf{A}_k(\mathbf{r}_P) &= \int_{-s_k/2}^{s_k/2} \frac{1}{|\mathbf{r}_P - \mathbf{r}_k(t)|}\frac{d\mathbf{r}_k}{dt}dt = \int_{-s_k/2}^{s_k/2} \frac{dt}{\sqrt{|\mathbf{r}_P - \mathbf{r}_k(t)|^2}}\,\hat{\mathbf{e}}_k \\ &= \ln\left\{\frac{-\left[\mathbf{r}_P - \mathbf{r}_k^+\right]\cdot\hat{\mathbf{e}}_k + \sqrt{\left|\mathbf{r}_P - \mathbf{r}_k^+\right|^2}}{-\left[\mathbf{r}_P - \mathbf{r}_k^-\right]\cdot\hat{\mathbf{e}}_k + \sqrt{\left|\mathbf{r}_P - \mathbf{r}_k^-\right|^2}}\right\}\hat{\mathbf{e}}_k\,, \end{aligned} \tag{5.8}$$

where

$$\begin{aligned}\mathbf{r}_k^{\pm} &= \mathbf{r}_k\left(\pm\tfrac{s_k}{2}\right) = \mathbf{R}_k \pm \tfrac{s_k}{2}\,\hat{\mathbf{e}}_k \\ &= \left[X_{0k} \pm \frac{s_k}{2}a_x\right]\hat{\mathbf{e}}_x + \left[Y_{0k} \pm \frac{s_k}{2}a_y\right]\hat{\mathbf{e}}_y + \left[Z_{0k} \pm \frac{s_k}{2}a_z\right]\hat{\mathbf{e}}_z, \quad (5.9)\end{aligned}$$

are the vector positions of the extremities of the current-carrying wire. Notice that a straight wire only produces a field $\mathbf{A}$ in a direction parallel to the wire. The field $\mathbf{B}$ can now be calculated from $\mathbf{B} = \nabla \times \mathbf{A}$. One may now combine any number of straight wires, possibly of different lengths, to form a closed figure. The fields for that case are obtained by an appropriate sum over each individual finite wire k.

There is one detail to point out. If $\left[\mathbf{r}_P - \mathbf{r}_k^{\pm}\right]\cdot\hat{\mathbf{e}}_k = |\mathbf{r}_P - \mathbf{r}_k^{\pm}|$, then Eq. (5.8) seems indeterminate[1]. This occurs if the point P in Fig. 5.2 lies along the line defined by $\hat{\mathbf{e}}_k$. For points along $\hat{\mathbf{e}}_k$ but outside the portion with the finite wire, the indetermination can be removed, and the result is

$$\frac{4\pi}{\mu_0 i}\mathbf{A}_k(\mathbf{r}_P) = \ln\left\{\frac{\left[\mathbf{r}_P - \mathbf{r}_k^{-}\right]\cdot\hat{\mathbf{e}}_k}{\left[\mathbf{r}_P - \mathbf{r}_k^{+}\right]\cdot\hat{\mathbf{e}}_k}\right\}\hat{\mathbf{e}}_k\,. \quad (5.10)$$

As expected, $\mathbf{A}_k(\mathbf{r}_P)$ diverges when the point P approaches the current-carrying wire.

5.3 Line integral along a straight path

5.3.1 General case

Following the spirit of Eq. (5.1), we need the integral of $\mathbf{A}_k$ along a piecewise linear closed loop. We start by calculating the integral of $\mathbf{A}_k$ along the straight line segment (designated by C_j) shown on the left in Fig. 5.3. This line is centered at

$$\mathbf{R}_j = X_{0j}\,\hat{\mathbf{e}}_x + Y_{0j}\,\hat{\mathbf{e}}_y + Z_{0j}\,\hat{\mathbf{e}}_z, \quad (5.11)$$

has the direction of the unit vector

$$\hat{\mathbf{e}}_j = b_x\hat{\mathbf{e}}_x + b_y\hat{\mathbf{e}}_y + b_z\hat{\mathbf{e}}_z, \quad (5.12)$$

[1] This may lead to problems in numerical simulations, which may be dealt with in one of two ways. One may program $\mathbf{A}_k(\mathbf{r}_P)$ and the ensuing calculations using Eqs. (5.8) and (5.10), or, alternatively, one may use only Eq. (5.8), introducing a very small numerical deviation in case of trouble. For example, in the two-square problem in Sec. 5.5, there are two wires on the same horizontal line when $\theta = 0 = \beta$ in Fig. 5.4. The numerical indetermination may be solved by choosing one square of side L and the other of side $(1+\delta)L$ with some very small δ (say, $\delta \sim 10^{-5}$).

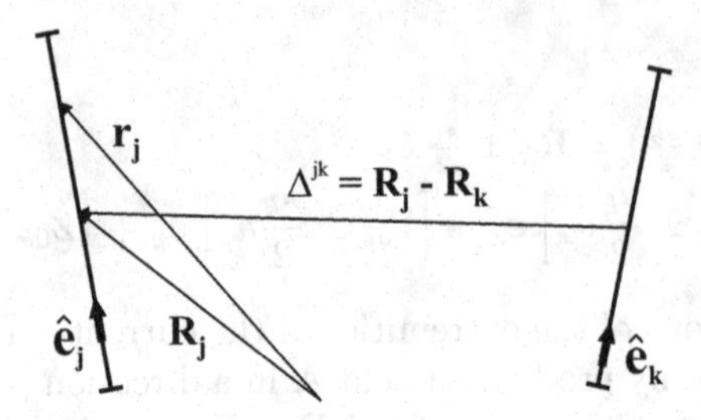

FIGURE 5.3
The finite straight circuit on the right carries a current i, creating the vector potential $\mathbf{A}_k$. This will be integrated along the straight line on the left in the figure. In general, the two lines are *not* on the same plane.

and length s_j. Any point along that line may be parametrized by

$$\begin{aligned}\mathbf{r}_j(t') &= \mathbf{R}_j + t'\,\hat{\mathbf{e}}_j & (5.13)\\ &= X_j(t')\,\hat{\mathbf{e}}_x + Y_j(t')\,\hat{\mathbf{e}}_y + Z_j(t')\,\hat{\mathbf{e}}_z & (5.14)\\ &= [X_{0j} + t'\,b_x]\,\hat{\mathbf{e}}_x + [Y_{0j} + t'\,b_y]\,\hat{\mathbf{e}}_y + [Z_{0j} + t'\,b_z]\,\hat{\mathbf{e}}_z, & (5.15)\end{aligned}$$

where $t' \in (-s_j/2, s_j/2)$.

The geometry of this problem depends on the direction of the current-carrying wire ($\hat{\mathbf{e}}_k$), on the direction of the integration path ($\hat{\mathbf{e}}_j$), and also on the vector connecting the centers of the two,

$$\mathbf{\Delta}^{jk} = \mathbf{R}_j - \mathbf{R}_k, \tag{5.16}$$

shown in Fig. 5.3. The calculations simplify considerably when the problem is effectively two-dimensional. This occurs in two cases. It can be that $\hat{\mathbf{e}}_k \neq \pm\hat{\mathbf{e}}_j$, but that the two segments lie on the same plane; this can be expressed by

$$\mathbf{\Delta}^{jk} \cdot (\hat{\mathbf{e}}_j \times \hat{\mathbf{e}}_k) = 0. \tag{5.17}$$

In this case, the relevant plane is defined by $\hat{\mathbf{e}}_j$ and $\hat{\mathbf{e}}_k$. On the other hand, the two lines might be parallel: $\hat{\mathbf{e}}_k = \pm\hat{\mathbf{e}}_j$. In that case, Eq. (5.17) is still true but the relevant plane is defined by $\hat{\mathbf{e}}_j$ and $\mathbf{\Delta}^{jk}$. If, however,

$$\mathbf{\Delta}^{jk} \cdot (\hat{\mathbf{e}}_j \times \hat{\mathbf{e}}_k) \neq 0, \tag{5.18}$$

then the problem is truly three-dimensional and, as we will see, the integration becomes more involved.

To calculate

$$\varphi_{jk} \equiv \int_{-s_j/2}^{s_j/2} \mathbf{A}_k\big|_{\mathbf{r}_p=\mathbf{r}_j} \cdot \frac{d\mathbf{r}_j}{dt'}\,dt', \tag{5.19}$$

we need

$$\frac{d\mathbf{r}_j}{dt'} = \hat{\mathbf{e}}_j. \tag{5.20}$$

Since $\mathbf{A}_k$ lies along $\hat{\mathbf{e}}_k$, the integral will be proportional to

$$\begin{aligned}\cos\alpha_{jk} &\equiv \hat{\mathbf{e}}_k \cdot \hat{\mathbf{e}}_j\\ &= a_x\,b_x + a_y\,b_y + a_z\,b_z, & (5.21)\end{aligned}$$

where α_{jk} is the angle between the straight current-carrying wire producing $\mathbf{A}_k$ and the integration path C_j.

Thus, using Eqs. (5.8) and (5.19)–(5.21), we concentrate on

$$\tilde{\varphi}_{jk} \equiv \frac{4\pi}{\mu_0 i} \frac{\varphi_{jk}}{\cos\alpha_{jk}}$$
$$= \int_{-s_j/2}^{s_j/2} \ln\left\{ \frac{-\left[\mathbf{r}_j(t') - \mathbf{r}_k^+\right]\cdot\hat{\mathbf{e}}_k + \sqrt{\left|\mathbf{r}_j(t') - \mathbf{r}_k^+\right|^2}}{-\left[\mathbf{r}_j(t') - \mathbf{r}_k^-\right]\cdot\hat{\mathbf{e}}_k + \sqrt{\left|\mathbf{r}_j(t') - \mathbf{r}_k^-\right|^2}} \right\} dt'. \tag{5.22}$$

We start by noting that the numerator inside the logarithm involves the distance between the forward extremity on the current-carrying wire and a generic point along the integration path:

$$\mathbf{r}_j(t') - \mathbf{r}_k^+ = (\mathbf{R}_j - \mathbf{r}_k^+) + t'\,\hat{\mathbf{e}}_j = \mathbf{\Delta}^{jk} - \frac{s_k}{2}\,\hat{\mathbf{e}}_k + t'\,\hat{\mathbf{e}}_j. \tag{5.23}$$

The projection of this vector onto the direction of the integration path $\hat{\mathbf{e}}_j$ is

$$v_+ = \left[\mathbf{r}_j(t') - \mathbf{r}_k^+\right]\cdot\hat{\mathbf{e}}_j = \mathbf{\Delta}^{jk}\cdot\hat{\mathbf{e}}_j - \frac{s_k}{2}\cos\alpha_{jk} + t', \tag{5.24}$$

leading to $dt' = dv_+$. We define two new parameters u_+ and β_{jk}^+, such that

$$\left|\mathbf{r}_j(t') - \mathbf{r}_k^+\right|^2 = u_+^2 + v_+^2 \tag{5.25}$$
$$-\hat{\mathbf{e}}_k\cdot\left[\mathbf{r}_j(t') - \mathbf{r}_k^+\right] = \sin\beta_{jk}^+\,u_+ - \cos\alpha_{jk}\,v_+. \tag{5.26}$$

The parameter α_{jk} and the variable v_+ have been defined already. From Eqs. (5.23)–(5.25) we obtain, after some algebra,

$$u_+^2 = \left|\mathbf{\Delta}^{jk}\right|^2 - \left(\mathbf{\Delta}^{jk}\cdot\hat{\mathbf{e}}_j\right)^2 + \frac{s_k^2}{4}\sin^2\alpha_{jk} - s_k\mathbf{\Delta}^{jk}\cdot\left(\hat{\mathbf{e}}_k - \cos\alpha_{jk}\hat{\mathbf{e}}_j\right) \tag{5.27}$$
$$= \frac{1}{\sin^2\alpha_{jk}}\left[\left|\mathbf{\Delta}^{jk}\cdot\left(\hat{\mathbf{e}}_k - \cos\alpha_{jk}\hat{\mathbf{e}}_j\right) - \frac{s_k}{2}\sin^2\alpha_{jk}\right|^2 + \left|\mathbf{\Delta}^{jk}\cdot\left(\hat{\mathbf{e}}_j\times\hat{\mathbf{e}}_k\right)\right|^2\right]. \tag{5.28}$$

Equation (5.28) holds when $\hat{\mathbf{e}}_k \neq \pm\hat{\mathbf{e}}_j$ ($\alpha_{jk} \neq 0, \pi$) [2]. We see that u_+ does *not* depend on the integration variable t' (i.e., it is the same for all points along the integration path). Now, from Eq. (5.26), we get

$$u_+\sin\beta_{jk}^+ = -\mathbf{\Delta}^{jk}\cdot\left(\hat{\mathbf{e}}_k - \cos\alpha_{jk}\hat{\mathbf{e}}_j\right) + \tfrac{s_k}{2}\sin^2\alpha_{jk}. \tag{5.30}$$

[2] Notice that, when $\hat{\mathbf{e}}_k \neq \pm\hat{\mathbf{e}}_j$, we can build an orthonormal basis with

$$\hat{\mathbf{e}}_a = \hat{\mathbf{e}}_j,$$
$$\hat{\mathbf{e}}_b = \frac{\hat{\mathbf{e}}_k - \cos\alpha_{jk}\,\hat{\mathbf{e}}_j}{\sin\alpha_{jk}},$$
$$\hat{\mathbf{e}}_c = \frac{\hat{\mathbf{e}}_j\times\hat{\mathbf{e}}_k}{\sin\alpha_{jk}}. \tag{5.29}$$

The combination $\hat{\mathbf{e}}_b$ appears in Eqs. (5.30), (5.45), and (5.46), and both $\hat{\mathbf{e}}_b$ and $\hat{\mathbf{e}}_c$ appear in Eq. (5.28).

This determines the sign of $u_+ \sin\beta^+_{jk}$, but not each sign individually. One possible definition for the signs is included in the appendix.

A similar analysis holds for the denominator inside the logarithm of Eq. (5.22), which involves the distance between the back extremity on the current-carrying wire and a generic point along the integration path: $\mathbf{r}_j(t') - \mathbf{r}^-_k$. This involves the new parameters

$$\begin{aligned} v_-(t') &= v_+(t') \text{ with } s_k \to -s_k, \\ u_- &= u_+ \text{ with } s_k \to -s_k, \\ \beta^-_{jk} &= \beta^+_{jk} \text{ with } s_k \to -s_k. \end{aligned} \tag{5.31}$$

To combine Eqs. (5.22), (5.25), and (5.26), we need

$$\begin{aligned} I(u,v) &\equiv \int \ln\left[u\sin\beta - v\cos\alpha + \sqrt{u^2+v^2}\right] dv \\ &= \begin{cases} I^{\alpha\neq\beta}(u,v) & \text{if } \alpha\neq\beta \\ I^{\alpha=\beta}(u,v) & \text{if } \alpha=\beta \end{cases}. \end{aligned} \tag{5.32}$$

The expressions for $I^{\alpha\neq\beta}(u,v)$ and $I^{\alpha=\beta}(u,v)$ are presented in the appendix.

Notice that, from Eqs. (5.28)–(5.30),

$$u^2\left(\sin^2\alpha_{jk} - \sin^2\beta_{jk}\right) = \left|\boldsymbol{\Delta}^{jk}\cdot(\hat{\mathbf{e}}_j\times\hat{\mathbf{e}}_k)\right|^2, \tag{5.33}$$

which vanishes if $\boldsymbol{\Delta}^{jk}\cdot(\hat{\mathbf{e}}_j\times\hat{\mathbf{e}}_k) = 0$. Thus, if the problem is effectively two dimensional, $I(u,v)$ reduces to the integral $I^{\alpha=\beta}(u,v)$—in Eq. (5.48) of the appendix.

Combining Eqs. (5.22), (5.25)–(5.26), and (5.32), we find

$$\frac{4\pi}{\mu_0 i}\varphi_{jk} = \cos\alpha_{jk}\left(I^+_{jk} - I^-_{jk}\right), \tag{5.34}$$

where

$$\begin{aligned} I^+_{jk} &= I_{jk}[u_+, v_+(t'=+\tfrac{s_j}{2})] - I_{jk}[u_+, v_+(t'=-\tfrac{s_j}{2})], & (5.35) \\ I^-_{jk} &= I_{jk}[u_-, v_-(t'=+\tfrac{s_j}{2})] - I_{jk}[u_-, v_-(t'=-\tfrac{s_j}{2})]. & (5.36) \end{aligned}$$

Eq. (5.35) depends on α_{jk} in Eq. (5.21), on $v_+(t'=\pm s_j/2)$ from Eq. (5.24), on u_+ from Eq. (5.45), and on β^+_{jk} from Eq. (5.46). Similarly, Eq. (5.36) depends on α_{jk} in Eq. (5.21), and on $v_-(t'=\pm s_j/2)$, u_-, and β^-_{jk}, defined in Eq. (5.31). Notice that, because u_+ does not involve the integration variable t', $u_+(+s_j/2) = u_+(-s_j/2) = u_+$, and, likewise, $u_-(+s_j/2) = u_-(-s_j/2) = u_-$.

5.3.2 Planar wires

Let us consider the possibility that the current-carrying wire and the integration path in Fig. 5.3 lie on the $x-y$ plane. In this case, $\hat{\mathbf{e}}_k$, $\hat{\mathbf{e}}_j$, and $\mathbf{\Delta}^{jk}$, lie on the same plane. Thus, $\mathbf{\Delta}^{jk}\cdot(\hat{\mathbf{e}}_j\times\hat{\mathbf{e}}_k)=0$, and we take $\beta^{\pm}_{jk}=\alpha_{jk}$. This means that, for planar cases, the $I(u,v)$ in Eqs. (5.34)–(5.36) are simply given by the $I^{\alpha=\beta}(u,v)$.

Writing

$$\begin{aligned}\hat{\mathbf{e}}_k &= -\sin\beta_k\,\hat{\mathbf{e}}_x+\cos\beta_k\,\hat{\mathbf{e}}_y,\\ \hat{\mathbf{e}}_j &= -\sin\gamma_j\,\hat{\mathbf{e}}_x+\cos\gamma_j\,\hat{\mathbf{e}}_y,\end{aligned} \tag{5.37}$$

we find

$$\begin{aligned}\cos\alpha_{jk} &= \hat{\mathbf{e}}_k\cdot\hat{\mathbf{e}}_j=\cos\beta_k\cos\gamma_j+\sin\beta_k\sin\gamma_j=\cos(\beta_k-\gamma_j),\\ \frac{\hat{\mathbf{e}}_k-\cos\alpha_{jk}\hat{\mathbf{e}}_j}{\sin\alpha_{jk}} &= -\left[\cos\gamma_j\,\hat{\mathbf{e}}_x+\sin\gamma_j\,\hat{\mathbf{e}}_y\right]\ \perp\ \hat{\mathbf{e}}_j,\end{aligned} \tag{5.38}$$

where we have defined $\alpha_{jk}=\beta_k-\gamma_j$.

As a result, Eqs. (5.34)–(5.36) hold, with the simplified variables:

$$u_+=(X_{0j}-X_{0k}+\frac{s_k}{2}\sin\beta_k)\cos\gamma_j+(Y_{0j}-Y_{0k}-\frac{s_k}{2}\cos\beta_k)\sin\gamma_j, \tag{5.39}$$

$$\begin{aligned}v_+(\pm\tfrac{s_j}{2}) &= -(X_{0j}-X_{0k}+\frac{s_k}{2}\sin\beta_k)\sin\gamma_j\\ &\quad+(Y_{0j}-Y_{0k}-\frac{s_k}{2}\cos\beta_k)\cos\gamma_j\pm\frac{s_j}{2},\end{aligned} \tag{5.40}$$

$$u_-=(X_{0j}-X_{0k}-\frac{s_k}{2}\sin\beta_k)\cos\gamma_j+(Y_{0j}-Y_{0k}+\frac{s_k}{2}\cos\beta_k)\sin\gamma_j, \tag{5.41}$$

$$\begin{aligned}v_-(\pm\tfrac{s_j}{2}) &= -(X_{0j}-X_{0k}-\frac{s_k}{2}\sin\beta_k)\sin\gamma_j\\ &\quad+(Y_{0j}-Y_{0k}+\frac{s_k}{2}\cos\beta_k)\cos\gamma_j\pm\frac{s_j}{2}.\end{aligned} \tag{5.42}$$

5.4 The magnetic flux and mutual inductance

Equations (5.34)–(5.36) are our new master equations. In that simple form, they can be used to describe the line integral of one current-carrying segment (k) along a straight segment (j). But they can also be used in more intricate problems. Consider a field generated by N_K line elements, each identified by a different $k=1...N_K$. The line integral of this field along a path made up of N_J straight paths can be obtained by summing Eq. (5.34) over k and j. If the latter form a closed integration path, then, due to Eq. (5.1),

$$\frac{4\pi}{\mu_0 i}\Phi=\frac{4\pi}{\mu_0 i}\sum_{j=1}^{N_J}\sum_{k=1}^{N_K}\varphi_{jk}=\frac{4\pi}{\mu_0 i}\sum_{j=1}^{N_J}\sum_{k=1}^{N_K}\cos\alpha_{jk}\left(I^+_{jk}-I^-_{jk}\right), \tag{5.43}$$

measures the magnetic flux that passes through the closed integration path.

Equations (5.34)–(5.36) generalize the equations obtained in Ref. [8], where the flux of a regular polygon of n sides was calculated across a smaller concentric polygon of n sides. Here: i) the length of each side is arbitrary; ii) we need not have the same number of sides in the current-carrying segments (N_K of them), and in the integration path (with N_J sides); iii) the circuits need not be concentric; and (most importantly); iv) the circuits need not even be planar, describing, instead, complicated forms in three dimensional space[3].

5.5 First application: two square wires on the plane

Let us now consider Fig. 5.4, showing two square wires of side L, whose centers are at a distance $r > \sqrt{2}L$ (to guarantee that the wires do not intersect). We

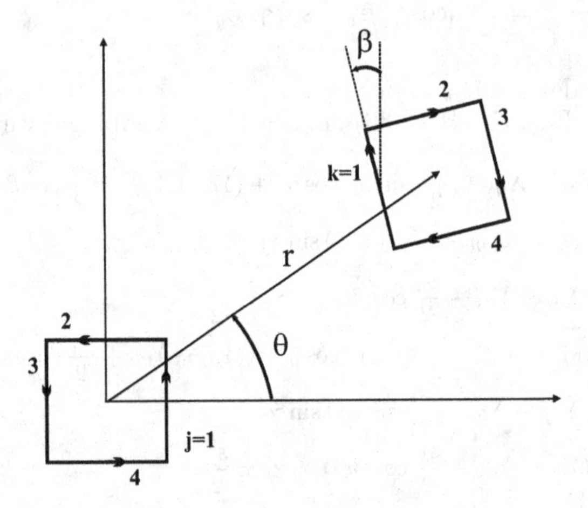

FIGURE 5.4
Two square wires of side L. The distance between the two centers is $r > \sqrt{2}L$, guaranteeing that the wires do not intersect. The wire on the right (with line segments denoted by k) carries a current i, which causes a magnetic flux on the square wire on the left (with line segments denoted by j).

have labeled the line segments on the wire on the right, starting from the left-most one ($k = 1$), and continuing clockwise: $k = 1 \ldots 4$. The parameters

[3] One can recover Eq. (18) of Ref. [8] with the substitutions, $N_J = N_K = n$, $\beta_k \to \beta_{nk}$, $X_{0k} \to d_n \cos\beta_{nk}$, $Y_{0k} \to d_n \sin\beta_{nk}$, $s_k \to s_n$, $\gamma_j \to 0$, $X_{0j} \to x_{\text{cut}}$, $Y_{0j} \to 0$, $s_j/2 \to x_{\text{cut}} \tan(\pi/n)$, and noting that all j terms are equal in that case, meaning that the sum over j yields simply an overall factor of n. The greatly simplifying substitution $\gamma_j \to 0$ appears because, in the case of two concentric polygons with n sides, the symmetry implies that the integration along any side is the same. Equations (5.39)–(5.42) are more complicated than the corresponding Eq. (15) in Ref. [8] precisely because here that simplification is absent.

needed for the determination of the **A** field produced by the current-carrying wire on the right of Fig. 5.4 are shown in Table 5.5. Henceforth, s_ϕ and c_ϕ

TABLE 5.1
Quantities relevant to the calculation of the **A** field created by the square loop on the right in Fig. 5.4.

k	β_k	$\hat{\mathbf{e}}_k$	X_{0k}	Y_{0k}
1	β	$-s_\beta\hat{\mathbf{e}}_x + c_\beta\hat{\mathbf{e}}_y$	$rc_\theta - \frac{L}{2}c_\beta$	$rs_\theta - \frac{L}{2}s_\beta$
2	$\beta - \frac{1}{2}\pi$	$c_\beta\hat{\mathbf{e}}_x + s_\beta\hat{\mathbf{e}}_y$	$rc_\theta - \frac{L}{2}s_\beta$	$rs_\theta + \frac{L}{2}c_\beta$
3	$\beta - \pi$	$s_\beta\hat{\mathbf{e}}_x - c_\beta\hat{\mathbf{e}}_y$	$rc_\theta + \frac{L}{2}c_\beta$	$rs_\theta + \frac{L}{2}s_\beta$
4	$\beta - \frac{3}{2}\pi$	$-c_\beta\hat{\mathbf{e}}_x - s_\beta\hat{\mathbf{e}}_y$	$rc_\theta + \frac{L}{2}s_\beta$	$rs_\theta - \frac{L}{2}c_\beta$

denote the sine and cosine of the angle ϕ, for any given angle ϕ.

Table 5.5 shows the parameters describing the rectangular integration path on the left of Fig. 5.4. We have labeled the right-most vertical segment with $j = 1$, continuing *anti-clockwise*: $j = 1 \ldots 4$.

TABLE 5.2
Quantities relevant for the calculation of the line integral along the path on the left in Fig. 5.4.

j	γ_j	$\hat{\mathbf{e}}_j$	X_{0j}	Y_{0j}
1	0	$\hat{\mathbf{e}}_y$	$\frac{L}{2}$	0
2	$\frac{1}{2}\pi$	$-\hat{\mathbf{e}}_x$	0	$\frac{L}{2}$
3	π	$-\hat{\mathbf{e}}_y$	$-\frac{L}{2}$	0
4	$\frac{3}{2}\pi$	$\hat{\mathbf{e}}_x$	0	$-\frac{L}{2}$

We can now substitute the values in Tables 5.5 and 5.5 into Eqs. (5.35)–(5.36) and (5.39)–(5.43). The results for the mutual induction are shown in Fig. 5.5 as a function of r, for $\theta = 0$ (displacement along the horizontal axis), and $\beta = 0, \pi/6, \pi/4$. As expected, for fixed θ and β, Φ decays with increasing distance. Henceforth, all distances are displayed in units of L and all mutual inductions are displayed in units of $\mu_0 iL/(4\pi)$. The mutual induction for $\beta = \pi/4$ is the largest. For values of $\beta > \pi/4$ we obtain the same curve as for $\pi/4 - \beta$, as one can immediately guess from Fig. 5.4. In particular, for $\beta = \pi/2$ we recover the curve for $\beta = 0$. Thus, for fixed relative position (θ and r), the mutual induction would in principle determine the relative orientation (β), up to discrete ambiguities. The problem, as we see from Fig. 5.5, is that the curve for $\beta = \pi/6$ lies so close to the curve for $\beta = \pi/4$, that it is barely visible. This means that it would take a very high precision to disentangle the two experimentally.

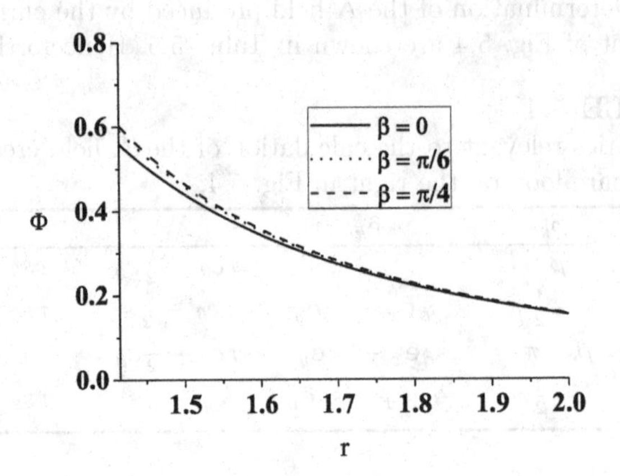

FIGURE 5.5
Mutual induction as a function of the distance between the loops in Fig. 5.4. The distance $r > \sqrt{2}$ is in units of L and Φ is in units of $\mu_0 i L/(4\pi)$. We have chosen $\theta = 0$ and: $\beta = 0$ (solid line); $\beta = \pi/6$ (dotted line); $\beta = \pi/4$ (dashed line).

A different situation occurs when we keep $\beta = 0$, and probe instead the variation with r for various values of θ. This is shown in Fig. 5.6, for $\theta = 0, \pi/6, \pi/4$. These curves also exhibit a symmetry $\theta \to \pi/4 - \theta$, with the highest values for Φ achieved for $\theta = \pi/4$. But the mutual inductance for $(\theta = \pi/4, \beta = 0)$ is larger than for $(\theta = 0, \beta = \pi/4)$. And the curve $\theta = \pi/6$ is easy to distinguish from the curve for $\theta = \pi/4$ in Fig. 5.6. The reason is merely geometrical. When $r \gtrsim \sqrt{2}L$, $\theta = 0$ and $\beta = \pi/4$, there is a gap between the two loops in Fig. 5.4. In contrast, when $r \gtrsim \sqrt{2}L$, $\theta = \pi/4$ and $\beta = 0$, the two loops almost touch. Thus, in the latter situation, they exhibit a larger mutual induction, explaining the difference between Figs. 5.5 and 5.6.

5.6 Second application: two square wires stacked

Let us now consider the situation depicted in Fig. 5.7, showing two square wires of side L, whose centers lie on the z axis, a distance h apart, with the square at $z = h$ rotated around the z axis through the angle β. The top (bottom) square lies on the $z = h$ ($z = 0$) plane. We can set $r = 0$ in Table 5.5, using it for $\hat{\mathbf{e}}_k$ (noting that it has no component along $\hat{\mathbf{e}}_z$), for X_{0k}, and for Y_{0k}, adding $Z_{0k} = h$ for all sides k. The j labeling in Fig. 5.7 differs from that in Fig. 5.4. We now get the parametrization shown in Table 5.6.

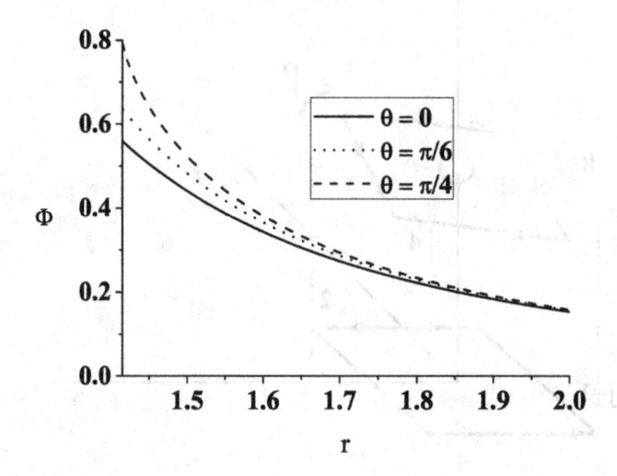

FIGURE 5.6
Mutual induction as a function of the distance between the loops in Fig. 5.4. The distance $r > \sqrt{2}$ is in units of L and Φ is in units of $\mu_0 i L/(4\pi)$. We have chosen $\beta = 0$ and: $\theta = 0$ (solid line); $\theta = \pi/6$ (dotted line); $\theta = \pi/4$ (dashed line).

Notice that, for $\beta \neq 0$, the problem is irreducibly three dimensional. Indeed, considering, for example, $j = k = 1$, we find that $\hat{\mathbf{e}}_{j=1} \times \hat{\mathbf{e}}_{k=1} = s_\beta \hat{\mathbf{e}}_z$, while

$$\boldsymbol{\Delta}^{11} = \frac{L}{2}(c_\beta - 1)\,\hat{\mathbf{e}}_x + \frac{L}{2} s_\beta\,\hat{\mathbf{e}}_y - h\,\hat{\mathbf{e}}_z. \tag{5.44}$$

Thus $\boldsymbol{\Delta}^{11} \cdot (\hat{\mathbf{e}}_{j=1} \times \hat{\mathbf{e}}_{k=1}) = -h s_\beta \neq 0$. This shows that even a problem as simple as the one presented in Fig. 5.7 requires the full three dimensional analysis, presented for the first time in this article.

Figure 5.8 shows the results for the mutual induction for $h = L/10$ and for $h = L/4$ as a function of the relative orientation angle β. The plots have the $\beta \to \pi/4 - \beta$ symmetry expected from Fig. 5.7. At $\beta = 0, \pi$ there is the maximum overlap between the loops, and the mutual induction takes its maximum value.

Let us first concentrate on the $h = L/10$ curve. The sharp drop when β increases from 0 has a very interesting physical origin. Consider the vertical component of the magnetic field **B** produced by the lower loop in Fig. 5.7. One can show that it has very sharp maxima close to the inside of the corners, and very sharp minima on the outside and close to the center of the sides of the square loop. As we rotate the upper square, we are rapidly moving away from the corners, where the maxima are, and moving into the influence of the sides, where the minima are. These two effects cause the sharp drop seen in the $h = L/10$ curve of Fig. 5.8; the maximum and minimum of that curve differ by about 30%.

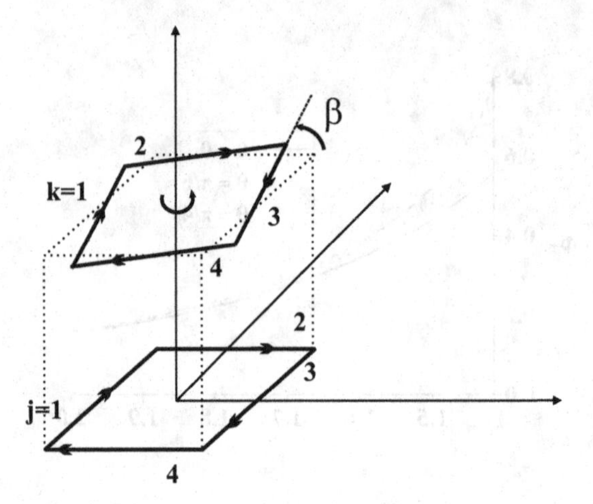

FIGURE 5.7
Two square wires of side L. The distance between the two centers is h (vertical). The lower (higher) square lies on the $z = 0$ ($z = h$) plane. The wire on top (with sides labeled by k), rotated by an angle β while remaining on the $z = h$ plane, carries a current i, which causes a magnetic flux on the square wire on the bottom (with line segments denoted by j). The labeling of the sides for both squares starts with 1 for the left-most side and continues clockwise.

Naturally, as h increases, the interaction is reduced and so is the mutual induction. But one notices two things in the $h = L/4$ curve of Fig. 5.8. First, the mutual induction is noticeably smaller. Second, and most importantly, the drop as β increases is not as sharp as before; the maximum and minimum of that curve differ only by about 8%. A curve for $h = L$ would have almost no variation at all. The reason is that the maxima and minima of $\mathbf{B}$ decrease very rapidly as h increases. They disappear almost completely when $h \sim L$. In fact, the field of the square approaches very rapidly that of a circle loop of equal radius and current. It is known that this should hold at large distances. What is interesting, and was first pointed out by Grivich and Jackson in Ref. [5], is that the field of the square is remarkably similar to that of the circle already at distances around $h \sim L$. It is this effect that makes the dependence of the mutual induction on β to become almost unnoticeable for $h > L$.

TABLE 5.3
Quantities relevant to parametrize the square wire on the lower ($z = 0$) plane of Fig. 5.7.

j	$\hat{\mathbf{e}}_j$	X_{0j}	Y_{0j}	Z_{0j}
1	$\hat{\mathbf{e}}_y$	$-\frac{L}{2}$	0	0
2	$\hat{\mathbf{e}}_x$	0	$\frac{L}{2}$	0
3	$-\hat{\mathbf{e}}_y$	$\frac{L}{2}$	0	0
4	$-\hat{\mathbf{e}}_x$	0	$-\frac{L}{2}$	0

5.7 Conclusions

Motivated by the simple calculation of the field produced by a finite line current, we have studied the mutual induction between two piecewise linear loops. We have calculated the vector potential **A** because its line integral yields the magnetic fluxes in a much simpler way. Our final result describes the mutual induction between two non-intersecting loops made of any number of straight lines, of arbitrary sizes and directions. As simple illustrations, we have studied the mutual influence between two square loops which lie on the same plane or which hover directly above each other. Our results can also be used to obtain approximations to the mutual induction of wires of arbitrary shape, by decomposing such wires into piecewise linear paths. The expressions presented here may also be of use in technological applications such as MEMS [9], or for estimating sensitivities in any imaging techniques where coils are used to obtain three-dimensional images from electromagnetic field distributions in the human body or in materials [10, 11].

Appendix

This appendix contains some calculational details which would detract from the flow of the text.

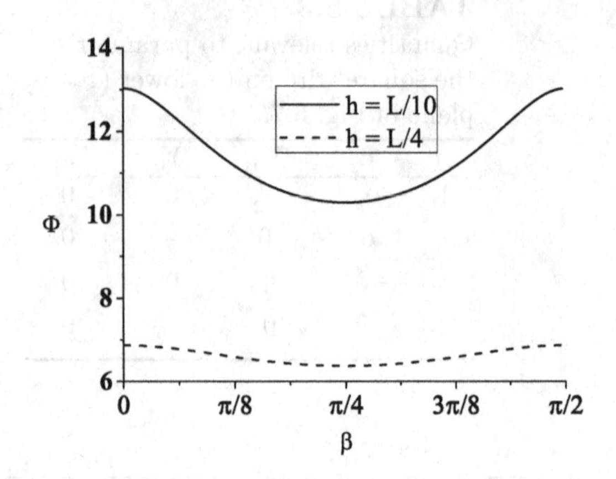

FIGURE 5.8
Mutual induction as a function of the relative orientation angle β, for the loops in Fig. 5.7. The distance h is in units of L, Φ is in units of $\mu_0 i L/(4\pi)$, and β is in radian. The curves correspond to $h = L/10$ (solid) and $h = L/4$ (dashed).

We have pointed out that Eq. (5.30) determines the sign of $u_+ \sin\beta^+_{jk}$, but not the signs of u_+ and $\sin\beta^+_{jk}$ individually. One possible definition is

$$u_+ = \begin{cases} \sqrt{|\boldsymbol{\Delta}^{jk}|^2 - (\boldsymbol{\Delta}^{jk} \cdot \hat{\mathbf{e}}_j)^2} & \text{if } \alpha_{jk} = 0, \pi \\ \frac{-\boldsymbol{\Delta}^{jk} \cdot (\hat{\mathbf{e}}_k - \cos\alpha_{jk} \hat{\mathbf{e}}_j) + \frac{s_k}{2} \sin^2\alpha_{jk}}{\sin\alpha_{jk}} & \text{if } \boldsymbol{\Delta}^{jk} \cdot (\hat{\mathbf{e}}_j \times \hat{\mathbf{e}}_k) = 0 \text{ and } \alpha_{jk} \neq 0, \pi \\ +\sqrt{u_+^2} & \text{if } \boldsymbol{\Delta}^{jk} \cdot (\hat{\mathbf{e}}_j \times \hat{\mathbf{e}}_k) \neq 0 \end{cases} \tag{5.45}$$

and, correspondingly,

$$\sin\beta^+_{jk} = \begin{cases} \sin\alpha_{jk} & \text{if } \boldsymbol{\Delta}^{jk} \cdot (\hat{\mathbf{e}}_j \times \hat{\mathbf{e}}_k) = 0 \\ \frac{-\boldsymbol{\Delta}^{jk} \cdot (\hat{\mathbf{e}}_k - \cos\alpha_{jk} \hat{\mathbf{e}}_j) + \frac{s_k}{2} \sin^2\alpha_{jk}}{u_+} & \text{if } \boldsymbol{\Delta}^{jk} \cdot (\hat{\mathbf{e}}_j \times \hat{\mathbf{e}}_k) \neq 0 \end{cases} . \tag{5.46}$$

As mentioned, when $\boldsymbol{\Delta}^{jk} \cdot (\hat{\mathbf{e}}_j \times \hat{\mathbf{e}}_k) = 0$, the problem is effectively two dimensional. The definitions in Eqs. (5.45) and (5.46) guarantee that, in that limit, $\sin\beta^+_{jk} = +\sin\alpha_{jk}$, and we can take $\beta^+_{jk} = \alpha_{jk}$.

We now turn to Eq. (5.32). We find

$$\begin{aligned}
I^{\alpha\neq\beta}(u,v) &= -v + u\csc^2\alpha\sin\beta\ln\left(v+\sqrt{u^2+v^2}\right)\\
&\quad +(v+u\csc\alpha\sin\beta\cot\alpha)\ln\left(u\sin\beta - v\cos\alpha+\sqrt{u^2+v^2}\right)\\
&\quad +2\,u\csc^2\alpha\sqrt{\sin^2\alpha-\sin^2\beta}\\
&\quad \times\arctan\left[\frac{(1-\sin\beta)\left(\sqrt{u^2+v^2}-u\right)-v\cos\alpha}{v\sqrt{\sin^2\alpha-\sin^2\beta}}\right], \qquad (5.47)
\end{aligned}$$

and

$$I^{\alpha=\beta}(u,v) \equiv \int \ln\left[u\sin\alpha - v\cos\alpha+\sqrt{u^2+v^2}\right]\,dv$$

$$= \begin{cases} v\,\ln\left(-v+\sqrt{u^2+v^2}\right)+\sqrt{u^2+v^2} & \text{if } \alpha=0\\ v\,\ln\left(v+\sqrt{u^2+v^2}\right)-\sqrt{u^2+v^2} & \text{if } \alpha=\pi\\ -v+u\,\csc\alpha\,\ln\left(v+\sqrt{u^2+v^2}\right) & \\ \quad +(v+u\cot\alpha)\ln\left(u\sin\alpha-v\cos\alpha+\sqrt{u^2+v^2}\right) & \text{otherwise}. \end{cases} \qquad (5.48)$$

One can check that these expressions are correct by performing the derivative with respect to v to recover Eq. (5.32).

Acknowledgments

We are very grateful to Augusto Barroso and to António J. Silvestre for reading and commenting on this manuscript. The research of A. C. Barroso was partially supported by Fundação para a Ciência e Tecnologia (Portuguese Foundation for Science and Technology) through the project PEst OE/MAT/UI0209/2011.

References

[1] See, for example, D. Halliday, R. Resnick, and J. Walker, *Fundamentals of Physics*, extended 6th ed. (John Wiley and Sons, New York, 2001).

[2] See, for example, D. J. Griffiths, *Introduction to Electrodynamics*, 3rd ed. (Prentice Hall, NJ, 1999).

[3] See, for example, J. D. Jackson, *Classical Electrodynamics*, 2nd ed. (John Wiley and Sons, New York, 1975).

[4] J. A. Miranda, "Magnetic field calculation for arbitrarily shaped planar wires," Am. J. Phys. **68**, 254–258 (1999).

[5] M. I. Grivich and D. P. Jackson, "The magnetic field of current-carrying polygons: An application of vector field rotations," Am. J. Phys. **68**, 469–474 (1999).

[6] H. Erlichson, "The magnetic field of a circular turn," Am. J. Phys. **57**, 607–610 (1989).

[7] H. G. Gnanatilaka and P. C. B. Fernando, "An investigation of the magnetic field in the plane of a circular current loop," Am. J. Phys. **55**, 341–344 (1987).

[8] J. P. Silva and A. J. Silvestre, "Comparing a current-carrying circular wire with polygons of equal perimeter: magnetic field versus magnetic flux," Eur. J. Phys. **26**, 783–790 (2005).

[9] A. Santra, N. Chakraborty, and R. Ganguly, "Analytical evaluation of magnetic field by planar micro-electromagnet spirals for MEMS applications," J. Micromech. Microeng. **19**, 085018 (2009).

[10] A. Gonzalez-Nakazawa, W. Q. Yang, and K. Hennessey, "An analytical approach for modelling electro-magnetic tomography sensor," Sensor Review **28**, 212–221 (2008).

[11] S. Ramani and J. A. Fessler, "Parallel MR image reconstruction using augmented Lagrangian methods," IEEE Transactions on Medical Imaging **30**, 694–706 (2011).

6

The Hertz Contact in Chain Elastic Collisions

P. Patrício

CONTENTS

A theoretical analysis of the influence of the Hertz elastic contact on a three body chain collision is presented. Despite the elastic character of the collision, the final velocity of each particle depends on the interaction between them. Two elastic spheres falling together, one on top of the other under the action of gravity, and then colliding with the ground, are studied in detail.

6.1 Introduction

When two balls are dropped together, both vertically aligned and hitting the ground at the same time, an interesting rebound effect is obtained. If the lighter ball is above, it may rise surprisingly high into the air. This effect may

Reproduced from P. Patrício, "The Hertz Contact in Chain Elastic Collisions," American Journal of Physics **72**, 1488–1491 (2004), https://doi.org/10.1119/1.1778394, with the permission of the American Association of Physics Teachers.

DOI: 10.1201/9781003187103-6

be enhanced if one drops a number of balls, ordered by their relative weight, the lightest ball at the top.

This effect was discussed more than 30 years ago, assuming perfect and independent elastic collisions between each pair of balls [1–5]. It was considered an attractive and ideal system to study elastic collisions. However, if the balls are dropped together, all collisions will happen at the same time. Because the collisions are not instantaneous, the final velocities of the balls will depend on the particular interactions between them, even in those cases where the interactions are elastic. Noninstantaneous collisions were considered in Ref. [2] by introducing a simple quasilinear model, in which the interaction between the balls was modeled by an effective harmonic spring. However, they concluded that the quasilinear model did not agree with their experimental results, and they instead proposed a phenomenological nonlinear model. The energy of the interaction was obtained directly from empirical data by dropping a ball onto a painted flat metal surface from varying heights. The radius of the paint spot on the surface of the ball was recorded. In this way the interaction was plotted as a function of the spot size, and eventually as a function of the depression of the ball. Although this strategy of obtaining the nonlinear energy for the interaction was remarkable, the analysis was not sufficiently developed. In particular, there was no attempt to point out the main effects of considering noninstantaneous collisions. The analysis of the noninstantaneous multiple collisions should take into account the interaction between two elastic spheres obeying Hooke's law, which is always valid for sufficiently small deformations. This interaction is known as the Hertz contact [6]. In this article, the multiple collisions of two elastic spheres with the same initial velocity are considered after the spheres hit the ground.

6.2 Independent collisions

Consider a system with two elastic spheres, one on top of the other, falling with the same speed v as represented in Fig. 6.1. Assume a very small gap between the spheres, such that all collisions can be treated individually. First, the lower sphere hits the ground and returns with the same speed v. Next, the spheres collide elastically. The final speeds u_1 and u_2 (of the lower and upper sphere, respectively) can be calculated using momentum and energy conservation, and are independent of the particular interaction. If $m = m_2/m_1$ is the spheres' mass ratio, we can write

$$u_1 = \frac{1-3m}{1+m}v, \tag{6.1}$$

$$u_2 = \frac{3-m}{1+m}v. \tag{6.2}$$

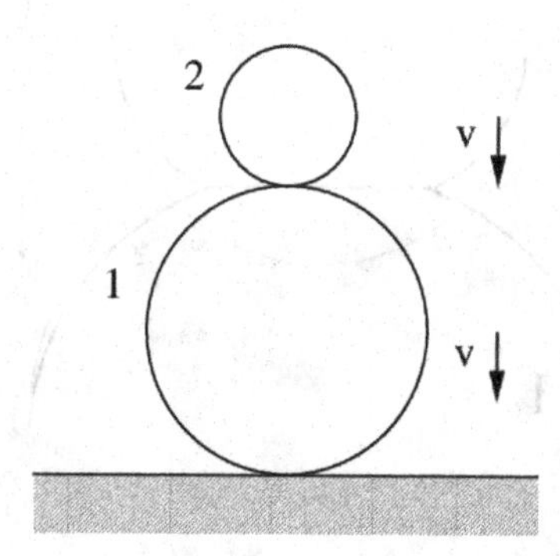

FIGURE 6.1
Two elastic spheres falling with the same velocity colliding with the ground.

The upper sphere (No. 2) may rebound with a speed up to three times greater than its initial one, for a vanishingly small value of m, the mass ratio.

The rebound may be enhanced if we add to this system extra spheres, all falling with the same initial speed v. To study this effect, consider first the elastic collision between two spheres with mass ratio m, but with different initial velocities. The upper sphere still moves with a downward speed v. The lower sphere has now previously collided with another sphere and moves upward with a speed av ($a > 1$). The final speeds are now given by

$$u_1 = \frac{a - (a+2)m}{1+m} v, \tag{6.3}$$

$$u_2 = \frac{1 + 2a - m}{1+m} v. \tag{6.4}$$

Equations (6.1) and (6.2) are recovered if $a = 1$. For a system with n spheres, the upper one may achieve a final speed of $(2n - 1)v$, if all mass ratios are vanishingly small.

Another interesting limit is the case for which all lower spheres stop and only the upper one rises into the air. In this case all the initial energy is transmitted to only one sphere. In the simple system composed of only two spheres, the lower one has zero final speed for $m = 1/3$, and the upper sphere with twice its initial speed.

6.3 Noninstantaneous collisions

If the spheres fall together, the assumption of independent collisions may no longer be accurate. For the system represented in Fig. 6.1, the collisions of the

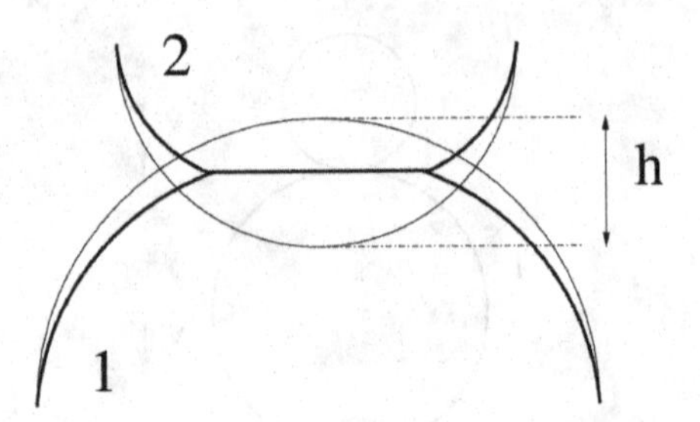

FIGURE 6.2
Elastic contact between two spheres.

lower sphere with the ground and between the two spheres will occur at the same time. To obtain the final speeds after the spheres separate, it is important to know the interaction between them. In the following, we present the simplest possible interaction: it will be assumed that all solids obey Hooke's law of elasticity, that is, that the stress-strain relation is linear. This law is valid for sufficiently small deformations. The dynamical equations for this model will then be established and solved.

6.3.1 The Hertz contact

The theory of the elastic contact between solids was first studied by Hertz and is discussed in detail in Ref. [6]. For two isotropic and homogeneous elastic spheres, with radii of curvature R_1 and R_2 (see Fig. 6.2), the elastic potential energy E_p depends on the combined deformation h as

$$E_p = \frac{2}{5} e r^{\frac{1}{2}} h^{\frac{5}{2}}, \tag{6.5}$$

where the reduced radius r is given by

$$r = \frac{R_1 R_2}{R_1 + R_2}, \tag{6.6}$$

and the reduced elastic constant e depends on the Young moduli E_1, E_2 and on the Poisson coefficients σ_1, σ_2:

$$e = \frac{4}{3}\left(\frac{1-\sigma_1^2}{E_1} + \frac{1-\sigma_2^2}{E_2}\right)^{-1}. \tag{6.7}$$

The Young's modulus increases with the rigidity, whereas the Poisson coefficient is typically slightly smaller than 1/2.

The elastic energy in Eq. (6.5) is valid for a static deformation. Nevertheless, it can be considered a good approximation for the case studied here if the velocity v is much smaller than the sound velocities of the solids involved. We note that even though the stress-strain relation is linear, due to geometrical reasons the energy in Eq. (6.5) increases as $h^{5/2}$. This approximation fits well the experimental potential energy obtained in Ref. [2] for small depressions of the balls. For large depressions, their experimental energy increased with a greater power of h, in agreement with the fact that Hooke's elastic regime in which the stress-strain linear relation is valid—takes into account only the first term of the elastic energy dependence on the deformation.

It is instructive to calculate the maximum extent of deformation and the contact time of two colliding spheres. Suppose ν is the relative velocity of the spheres before the collision, and $\mu = m_1 m_2/(m_1 + m_2)$ their reduced mass. Then, energy conservation gives

$$\frac{1}{2}\mu\left(\frac{dh}{dt}\right)^2 + \frac{2}{5}er^{\frac{1}{2}}h^{\frac{5}{2}} = \frac{1}{2}\mu\nu^2. \tag{6.8}$$

The maximum height of deformation occurs when the relative velocity is zero and can be written as

$$h_M = \left(\frac{5\mu}{4er^{\frac{1}{2}}}\right)^{\frac{2}{5}}\nu^{\frac{4}{5}}. \tag{6.9}$$

The collision time corresponds to the time in which the deformation goes from 0 to h_M and back to 0 again. It is given by

$$\tau = 2\left(\frac{25\mu^2}{16e^2 r\nu}\right)^{\frac{1}{5}}\int_0^1 \frac{dx}{\sqrt{1-x^{\frac{5}{2}}}} \approx 3.21\left(\frac{\mu^2}{e^2 r\nu}\right)^{\frac{1}{5}}. \tag{6.10}$$

These are the relevant length and time for elastic collision problems, and they will be useful for obtaining dimensionless dynamical equations.

6.3.2 Dynamical equations

For the system represented in Fig. 6.1, the total potential energy is given by

$$E_p = \frac{2}{5}e_{01}r_{01}^{\frac{1}{2}}h_{01}^{\frac{5}{2}} + \frac{2}{5}e_{12}r_{12}^{\frac{1}{2}}h_{12}^{\frac{5}{2}}, \tag{6.11}$$

The first term takes into account the interaction between the first sphere and the rigid ground (represented by the index No. 0), with infinite Young modulus and radius of curvature, and the second term is the interaction between the two spheres. The reduced radius r_{ij} and reduced elastic constant e_{ij} are given by Eqs. (6.6) and (6.7) ($ij = 01, 12$). The deformations depend on the positions x_1 and x_2 of the centers of the spheres (the ground defines the reference position)

$$h_{01} = H\left(R_1 - x_1\right), \tag{6.12}$$

$$h_{12} = H\left((R_1 + R_2) - (x_2 - x_1)\right). \tag{6.13}$$

The function $H(x) = x$ if $x > 0$; $H(x) = 0$ otherwise. The equations of motion $m_i\ddot{x}_i = -\partial E_p/\partial x_i, (i = 1, 2)$ are

$$m_1\ddot{x}_1 = e_{01}r_{01}^{\frac{1}{2}}h_{01}^{\frac{3}{2}} - e_{12}r_{12}^{\frac{1}{2}}h_{12}^{\frac{3}{2}} \tag{6.14}$$

$$m_2\ddot{x}_2 = e_{12}r_{12}^{\frac{1}{2}}h_{12}^{\frac{3}{2}}. \tag{6.15}$$

If we introduce natural units, $L = \left(m_1v^2/e_{01}r_{01}^{\frac{1}{2}}\right)^{\frac{2}{5}}$ and $T = L/v$, which are associated with the maximum height of depression and the time of contact between the ground and the first sphere, it is possible to define dimensionless variables. The equations of motion may be written in a simpler form

$$\ddot{x}_1 = h_{01}^{\frac{3}{2}} - kh_{12}^{\frac{3}{2}}, \tag{6.16}$$

$$m\ddot{x}_2 = kh_{12}^{\frac{3}{2}}, \tag{6.17}$$

with two parameters,

$$m = m_2/m_1, \tag{6.18}$$

$$k = e_{12}r_{12}^{\frac{1}{2}}/e_{01}r_{01}^{\frac{1}{2}}. \tag{6.19}$$

Note that k is defined as a ratio of the reduced elastic constants and radius. If the infinite rigidity of the ground is taken into account, we have $e_{01} > e_{12}$ and $r_{01} > r_{12}$. For this system, we conclude that $k < 1$.

6.3.3 Numerical resolution

The set of differential dynamical Eqs. (6.16) and (6.17) was solved numerically using the simple Euler algorithm for different values of the parameters. Figure 6.3 shows several solutions for x_1 and x_2 as functions of time, starting when the two spheres initially in contact hit the ground with the same velocity v, until they separate with constant final velocities u_1 and u_2. The axes have arbitrary units. The solutions were calculated for spheres with the same Young modulus and density, but four different mass ratios $m = 0.01, 0.5, 1, 5$ (which were obtained for appropriate sphere radii).

For a very small mass ratio, the spheres have almost the same velocity $u_1 \approx u_2 \approx v$ after the rebound. This situation contrasts with the results obtained assuming independent collisions, for which the maximum velocity gain would be expected. If the mass ratio m is negligible, we can see from Eq. (6.17) that the deformation h_{12} also will be very small, and will have little influence on the motion of the lower sphere (see Eq. (6.16)). The latter sphere rebounds from the ground and returns with velocity $u_1 \approx v$. Since $h_{12} \approx R_2 \approx 0$, we conclude that $R_1 \approx x_2 - x_1$. Then $\ddot{x}_1 \approx \ddot{x}_2$, which means that both spheres stick together during the collision as if they were one body [1].

[1] This argument was first suggested by J. P. Silva

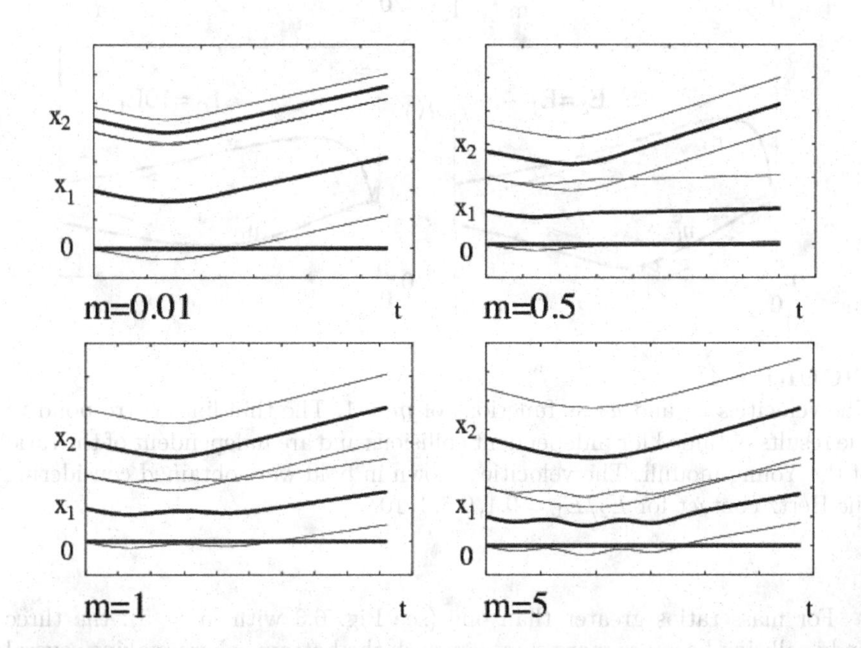

FIGURE 6.3
The positions x_1 and x_2 as a function of time, for spheres with the same Young modulus and density, and mass ratios $m = 0.01, 0.5, 1, 5$. The axes have arbitrary units. The bold lines represent the ground and the center of the spheres. The thin lines represent $x_1 \pm R_1$ and $x_2 \pm R_2$ and indicate at which times the interaction exists.

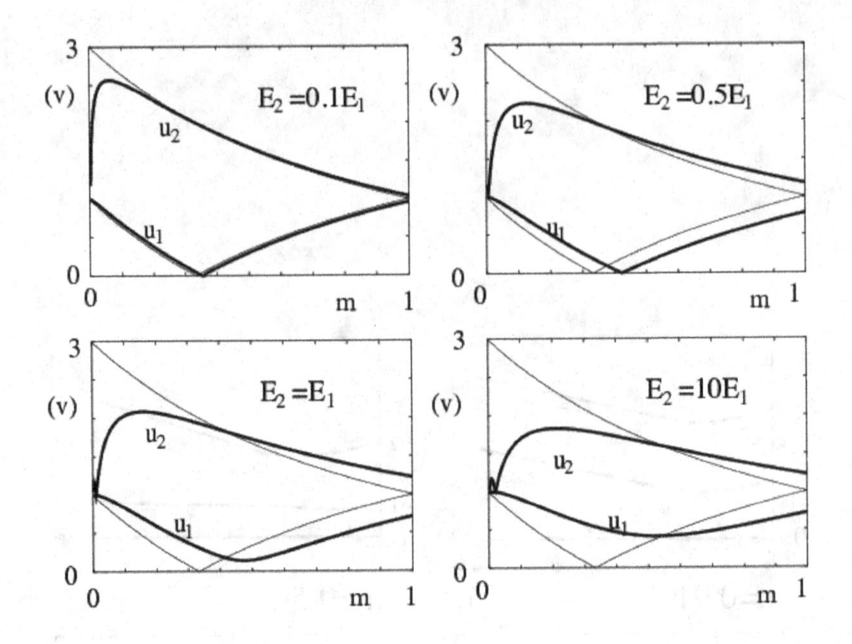

FIGURE 6.4
The velocities u_1 and u_2 as functions of $m < 1$. The thin lines correspond to the results obtained for independent collisions and are independent of the ratio of the Young moduli. The velocities shown in bold were obtained considering the Hertz contact for $E_2/E_1 = 0.1, 0.5, 1, 10$.

For mass ratios greater than one (see Fig. 6.3 with $m = 5$), the three body collision becomes more complex, with the bottom sphere making several rebounds between the ground and the other sphere. Note that the final velocities do not depend on the amplitude of the spheres' deformations during the collisions. If both spheres were more rigid, but with the same ratio k, the deformations could be very small, but the final velocities would still be equal to the ones shown in Fig. 6.3.

The independent collision results are recovered for small values of E_2/E_1, and consequently for small k. This observation is particularly true for the final velocity of the lower sphere u_1. If $k \to 0$, we can see from Eq. (6.16) that immediately after the collision, the motion of the lower sphere will depend mainly on its interaction with the ground—the deformation h_{12} between the two spheres will have a negligible contribution compared to the deformation h_{01} between the lower sphere and the ground. However, if the collision time between the two spheres is large enough (in Eq. (6.10), the collision time increases for small k), after the lower sphere-ground interaction, the terms containing the deformation h_{12} will be the only ones present in the equations

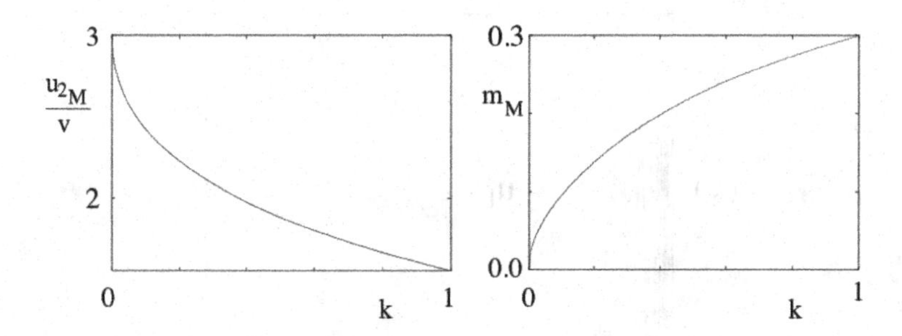

FIGURE 6.5
Maximum rebound speed u_{2M} and corresponding mass ratio m_M as a function of $k < 1$ (assuming that the ground is infinitely rigid).

of motion. Therefore, the collisions can be considered as almost independent and consecutive. Nevertheless, the convergence does not occur for all values of m; it is possible to see in Fig. 6.4 ($E_2/E_1 = 0.1$) that the final speed $u_2 \to v$ as the mass ratio $m \to 0$.

As E_2/E_1 (or k) increases, small changes can be seen in the final velocities, and the independent collision approximation is no longer applicable. These changes are limited because the infinite rigidity assumption requires that $k < 1$. The velocities for $E_2/E_1 \gg 1$ are not very different from the one for $E_2/E_1 = 10$.

The left part of Fig. 6.5 shows the rebound maximum speed u_{2M} as a function of $k < 1$. On the right, the mass ratio m_M corresponding to the maximum rebound speed is also plotted. m_M decreases rapidly to zero as $k \to 0$.

The discontinuity of the slope observed for the final velocity u_1 in Fig. 6.5 ($E_2/E_1 = 0.1$) corresponds to the critical value m_c for which the lower sphere stops completely. If $m > m_c$, this sphere rebounds twice off the ground, before it reaches the final velocity shown in the figure. If E_2/E_1 increases, the discontinuity changes first its position ($E_2/E_1 = 0.5$), and eventually it disappears ($E_2/E_1 = 1$). However, these discontinuities will occur again for larger values of m, as the lower sphere rebounds more and more in between the ground and the upper sphere. This curious feature can be seen in Fig. 6.6, which shows the velocities u_1 and u_2 as a function of $1/m < 1$, for $E_2/E_1 = 1$. As $1/m$ approaches 0, that is, when the lower sphere becomes much lighter than the top one, the number of collisions increases to infinity. In a recent article, Redner mapped the problem of three arbitrary particles colliding elastically and instantaneously into an equivalent billiard system [7]. From this approach, the entire particle collision history can be inferred in a very simple geometric manner. For the particular system considered here (in the limiting case of independent collisions), the maximum possible number of

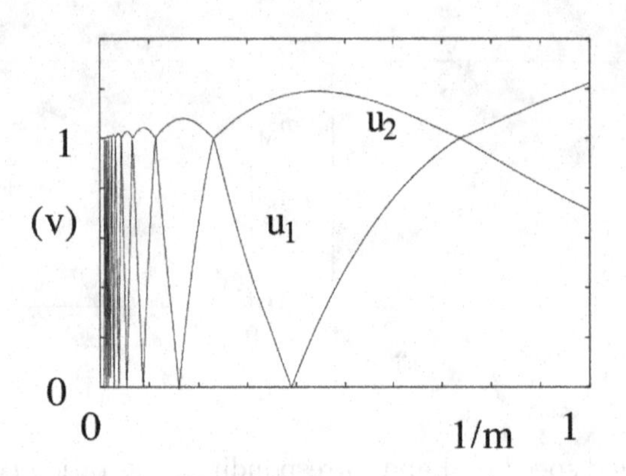

FIGURE 6.6
The velocities u_1 and u_2 as functions of $1/m < 1$, for $E_2/E_1 = 1$.

collisions N_M for large m was found to be

$$N_M \approx \pi\sqrt{m}. \tag{6.20}$$

6.4 Discussion and conclusions

The influence of the Hertz contact on multiple chain collisions was discussed in detail, and it was shown that the independent collision approximation is no longer valid for small values of the mass ratio m and large values of the rigidity ratio k. The results obtained with this model are summarized in Figs. 6.4 and 6.5, which clearly indicate the best parameters m and k to choose experimentally to see the influence of the Hertz contact on chain elastic collisions.

To understand our results, it is important to consider the typical two-body collision times (see Eq. (6.10)). First, suppose m is of order unity. If k is small, the two-body collision time between the lower sphere and the ground, τ_{01}, would be smaller than the two-body collision time between the two spheres, τ_{12}. As $k \to 0$, the first collision occurs almost instantaneously, and subsequently, the two spheres collide. Thus, the results obtained in this limit approach the independent collision results. However, as k increases, τ_{01} also increases and it is not possible to treat the collisions separately.

For large k and small m, the final velocities u_1 and u_2 approach each other. In this limit ($m \to 0$), the two spheres behave as if they belong to the same body, changing their velocities together as they interact with the ground. At the end, both spheres rebound with equal final velocities $u_1 \approx u_2 \approx v$.

Acknowledgments

The author would like to thank J. P. Silva for getting him interested in the multiple chain collision problem, for very useful discussions, and for a critical reading of the article. The author also thanks J. M. Tavares for his enthusiasm and for enlightening comments.

References

[1] W. R. Mellen, "Superball rebound projectiles," Am. J. Phys. **36**, 845 (1968).

[2] Class of W. G. Harter, "Velocity amplification in collision experiments involving superballs," Am. J. Phys. **39**, 656–663 (1971).

[3] J. D. Kerwin, "Velocity, momentum, and energy transmissions in chain collisions," Am. J. Phys. **40**, 1152–1157 (1972).

[4] J. S. Huebner and T. L. Smith, "Multi-ball collisions," Phys. Teach. **30**, 46–47 (1992).

[5] W. R. Mellen, "Aligner for elastic collisions of dropped balls," Phys. Teach. **33**, 56–57 (1995).

[6] See, for instance, L. D. Landau and E. M. Lifshitz, *Theory of Elasticity* (Pergamon Press, Oxford, 1986).

[7] S. Redner, "A billiard-theoretic approach to elementary 1d elastic collisions," Am. J. Phys. **72**, 1492–1498 (2004).

7

Tilted Boxes on Inclined Planes

A. M. Nunes and J. P. Silva

CONTENTS

We propose the study of a box placed on an inclined plane, with an initial tilt with respect to the plane. This is a paradigmatic example of the role played by friction as a link between translational and rotational motion. This example has two advantages over the usual example of a sphere (or cylinder) rolling down an inclined plane. First, it provides a good model for a much greater variety of "real-life" situations. Second, it exhibits a much richer structure in parameter space, even when the box starts from rest.

Reproduced from A. M. Nunes and J. P. Silva, "Tilted boxes on inclined planes," American Journal of Physics **68**, 1042–1049 (2000), https://doi.org/10.1119/1.1286047, with the permission of the American Association of Physics Teachers.

DOI: 10.1201/9781003187103-7

7.1 Introduction

Friction plays a very important role in the study of mechanical models. Most students learn to appreciate how different the world would be, were it not for friction. How would we walk? How would we stop? Where would our furniture go if we leaned on it? As a result, all books of introductory physics contain chapters on this subject (see, for example, Refs. [1, 2]). Intermediate or advanced undergraduate level textbooks on Classical Mechanics usually include also topics on viscous friction, and on the role of friction in typical nonholonomic systems, such as the rolling coin (see, for example, Refs. [3, 4]).

One interesting topic concerns the crucial role played by friction as the source of the coupling between translational and rotational motion. The prototypic example of this occurs in the case of a sphere which rolls down an inclined plane, starting from rest. However, we have found it helpful to present the students with other examples highlighting this important role of friction. In fact, the example of a sphere which starts from rest and rolls down an inclined plane provides a poor representation for items of furniture, houses, or even people placed on slopes. In particular, it does not accept solutions in which the object slides down while rotating backwards (as happens often with skiers) or solutions in which the edge of the object slides upwards while the object rotates forward (an interesting case which could happen if the same skier were silly enough as to lean forward excessively). It is true that these solutions do show up in the case of a sphere placed on an inclined plane with nonzero initial velocity. But, that example by itself might induce the incorrect notion that such types of motion require an initial velocity. This is definitely not the case with boxes and most real-life objects.

In addition, the sphere will never stay motionless on an inclined plane, while most other objects may. This behavior is related with the line of action of the reaction of the plane (perpendicular to the plane).

Remarkably, a problem as simple as the motion of a box allowed to move and/or rotate, and placed initially at rest on an inclined plane exhibits a rich parameter-space structure which illustrates all these issues and may be used in order to highlight the role played by friction in some interesting real-life situations.

In Sec. 7.2 we concentrate on the case in which the box rests evenly on the inclined plane. In this case there are stable solutions in which the box does not rotate. We stress the role that the position of the normal force exerted by the plane on the box has on this state. Understanding this will lead the students to appreciate immediately why spheres cannot rest on inclined planes. They will also understand why this is also the case with boxes placed initially at a tilt with respect to the inclined plane. This case is discussed in detail in Sec. 7.3, where it is shown that solutions occur in which the edge of the box slides up (down) the inclined plane while the box rotates forward (backward). We draw our conclusions in Sec. 7.4.

7.2 Boxes resting evenly on the plane

Let us consider a uniform box of base b, height h an mass m, placed at rest on an inclined plane of angle θ. In this section we will consider a box placed evenly on the plane (i.e., with its base parallel to the plane), as shown in Fig. 7.1. In connection with Fig. 7.1, we refer to the counter-clockwise rotation as "tumbling forward" and to the clockwise rotation as "tumbling backward": it is as if the box were a person facing down the slope.

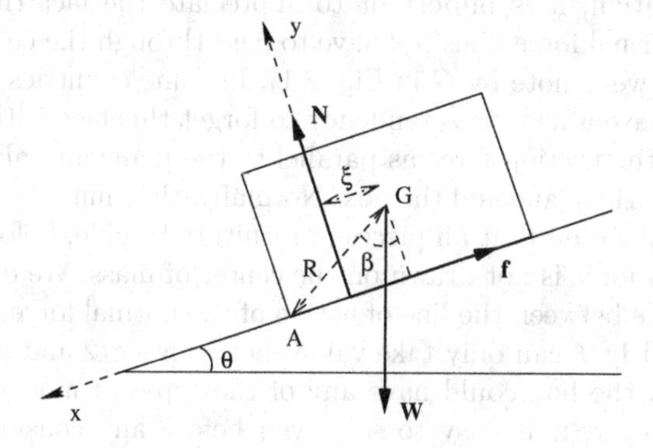

FIGURE 7.1
Forces acting on a box resting evenly on the plane. The box has base b and height h, which are related to R and β through Eqs. (7.1).

We use the coordinates x, parallel to the plane, and y, perpendicular to the plane, to describe the position of the lower edge of the box: x increases (decreases) as the lower edge slides down (up) the plane, while y remains constant (and we set it equal to zero). In addition, we will choose the counter-clockwise rotation (tumbling forward) as positive, corresponding to the choice of a left-handed coordinate system. We assume that the box is symmetric with respect to the plane of Fig. 7.1. Therefore, we will ignore the extra dimension and, in referring to this figure, we will talk about the line of contact between the box and the plane (*c.f.* the surface of contact between the box and the plane).

We may trade b and h for

$$\tan\beta = b/h,$$
$$R = \sqrt{(b/2)^2 + (h/2)^2}. \tag{7.1}$$

Clearly, $\tan\beta$ controls the shape of the box and is scale independent, while R sets the scale for the size of the box. Flat objects correspond to $\tan\beta \gg 1$; slim objects correspond to $\tan\beta \ll 1$. We denote by g the acceleration due to gravity, by W ($W = mg$) the weight of the box, by f the friction force, and by N the reaction of the plane on the box, which acts perpendicularly to the plane (henceforth, we shall call this the "normal force"). Strictly speaking, the last two forces are distributed all along the surface of contact between the box and the plane. Nevertheless, under the assumption of rigid bodies, we may substitute all the normal forces by their resultant, as long as we place it along a specific line of action, choosen in such a way as to reproduce the same torque as the sum (integral) of all the torques of the individual normal forces. It is this resultant that we denote by N. Similarly, we may substitute the distribution of frictional forces by a single force, parallel to the plane, which we denote by f.

In this situation, it is important to appreciate the fact that the line of action of the normal force does *not* have to pass through the center of mass of the box (which we denote by G in Fig. 7.1). In some countries, including our own, students have a notorious tendency to forget this fact, although they do draw correctly the friction force as parallel to the plane and along the line of contact between the plane and the box. Normally, they must be faced with the nonsensical implication that all pieces of furniture tumble, before they realize that the normal force is not drawn on the center of mass. We denote by ξ the oriented distance between the line of action of the normal force and the center of mass. Obviously, ξ can only take values between $-b/2$ and $+b/2$.

In principle, the box could have any of the types of motion described in Table 7.2. However, it is easy to see, even before any consideration of the dynamics, that the three cases UN, UB, and NB are not possible. In fact, any of these cases would increase the total energy of the box.

TABLE 7.1
Classification of the types of motion allowed.

	Tumble forward	No tumble	Tumble backward
Sliding down	DF	DN	DB
No sliding	NF	NN	NB
Sliding up	UF	UN	UB

Since the box starts out by resting evenly on the plane, dynamical considerations will also exclude the cases DB and UF. In fact, if the box slides down the plane, the friction force is drawn upwards (as in Fig. 7.1), inducing a counter-clockwise rotation on the box (with respect to G). On the other hand, in order to rotate backwards, the box would have to lean on its upper edge and, thus, this would be the only contact point with the surface. As a result, the normal would be drawn on that edge and would also induce a counter-clockwise rotation on the box (with respect to G). Therefore, the resulting torque on the box, with respect to G, would be counter-clockwise, while we have assumed that the box would tumble backward, i.e., clockwise. This contradiction eliminates the case DB from consideration. A similar analysis excludes the case UF.

We have concluded that the edge of a box which initially rests evenly on a plane, may either slide down or not slide at all; it will not slide up. Also, the box may rotate forward or not rotate at all; but it will not rotate backward.

The equations of motion of the box for $t = 0$ are

$$\begin{cases} N - mg\cos\theta &= m\alpha R\sin\beta \\ mg\sin\theta - f &= ma + m\alpha R\cos\beta \\ f\frac{h}{2} - N\xi &= m\eta R^2\alpha. \end{cases} \tag{7.2}$$

In writing these equations, we have used the fact that, since we start from rest, the acceleration of the center of mass (a_G) is related with the angular acceleration (α) and with the acceleration of the (lower) edge along the plane (a) by

$$\begin{aligned} (a_G)_x &= a + \alpha R\cos\beta, \\ (a_G)_y &= \alpha R\sin\beta. \end{aligned} \tag{7.3}$$

We recall that we have used a convention in which the counter-clockwise rotation and the downward translation are taken as positive. In addition, we have written the moment of inertia of the box about an axis perpendicular to the plane of Fig. 7.1 and passing through the center of mass as

$$I_G = \frac{1}{12}m\left(b^2 + h^2\right) = m\eta R^2, \tag{7.4}$$

meaning that $\eta = 1/3$.

We must now separate the case with no sliding ($a = 0$), where the friction is static, from the case in which the box will slide down the plane ($a > 0$), where the friction will become kinetic. In either case, the friction force is directed up the plane. Although this is not essential, we will take the naive approximation to friction in which the static and kinetic friction forces depend on N through

$$\begin{aligned} f_s &\leq f_s^{max} = \mu_s N, \\ f_k &= \mu_k N, \end{aligned} \tag{7.5}$$

respectively. The static and kinetic friction coefficients (μ_s and μ_k, respectively) are taken to be constants.

In this context, we should make the following remark. Strictly speaking, the kinetic friction should only be used once $\dot{x} \neq 0$, while we are assuming that the box starts from rest. In fact, the essential assumption in Eq. (7.3) is that the box starts from *rotational* rest, that is, that the initial rotational velocity is zero. Therefore, these equations will break down for $t > 0$ whenever the box tumbles. However, the backward time limit $t \to 0^+$ will be suitably described by our analysis, provided that we take $\mu = \mu_k$. Technically, this occurs because the friction coefficient as a function of time is discontinuous at $t = 0$, while the rotational velocity is a continuous function of time and vanishes, by assumption, at $t = 0$.

Morevover, Eqs. (7.2) and (7.3) still hold when we allow the possibility of a positive initial sliding velocity $\dot{x}(0) > 0$. This makes our analysis of the $a > 0$ cases valid both when $\dot{x}(0) = 0$ (and $\mu = \mu_s$), and when $\dot{x}(0) > 0$ (and $\mu = \mu_k$). Therefore, we will use the loose notation μ and f for the friction coefficient and force, respectively.

We will also have to separate the case in which there is no tumbling ($\alpha = 0$), where we must search for the position ξ of the line of action of the normal that guarantees that there is no tumbling, from the case in which the box tumbles forward ($\alpha > 0$), where the contact point will only be the lower edge and ξ takes necessarily the value $b/2$.

7.2.1 Case 1: no sliding and no tumbling

In this case $a = 0$, $\alpha = 0$ and $f_s \leq \mu_s N$. Equations (7.2) may be solved for the three unknowns N, f_s, and ξ. One finds,

$$
\begin{aligned}
N &= mg\cos\theta, && (7.6)\\
f_s &= mg\sin\theta, && (7.7)\\
\xi &= \frac{h}{2}\tan\theta. && (7.8)
\end{aligned}
$$

We must now impose the physical conditions. These lead to

$$
\begin{aligned}
\tan\theta &\leq \tan\beta,\\
\mu_s &\geq \tan\theta, && (7.9)
\end{aligned}
$$

where the first equation follows from $\xi \leq b/2$ and the second from $f_s \leq \mu_s N$. The requirement that $N \geq 0$ is trivially satisfied here, and also in cases 2 and 3.

7.2.2 Case 2: no sliding and tumbling forward

In this case $a = 0$, $\xi = b/2$, and $f_s \leq \mu_s N$. Equations (7.2) may be solved for the three unknowns N, f_s, and α. One finds,

$$
\begin{aligned}
\frac{N}{mg\cos\theta} &= \frac{1}{1+\eta}\left[\eta + \cos^2\beta + \sin\beta\,\cos\beta\,\tan\theta\right], && (7.10)\\
\frac{f_s}{mg\cos\theta} &= \frac{1}{1+\eta}\left[(\eta + \sin^2\beta)\tan\theta + \sin\beta\,\cos\beta\right] && (7.11)\\
\alpha &= \frac{g}{R}\frac{1}{1+\eta}\sin(\theta-\beta). && (7.12)
\end{aligned}
$$

The conditions $\alpha \geq 0$ and $f_s \leq \mu_s N$ lead to

$$\tan\theta \geq \tan\beta, \tag{7.13}$$

$$\mu_s \geq \frac{(\eta + \sin^2\beta)\tan\theta + \sin\beta\,\cos\beta}{\eta + \cos^2\beta + \sin\beta\,\cos\beta\,\tan\theta}, \tag{7.14}$$

respectively. The equality in Eq. (7.14) represents an hyperbola in the $(\tan\theta, \mu_s)$ plane, centered at

$$(\tan\theta, \mu_s) = \left(-\frac{\eta}{\sin\beta\,\cos\beta} - \cot\beta, \frac{\eta}{\sin\beta\,\cos\beta} + \tan\beta\right) \tag{7.15}$$

and with asymptotes parallel to the coordinate axis. In the first quadrant of the $(\tan\theta, \mu_s)$ plane (our physical region) this curve starts out at $(\tan\theta, \mu_s) = (\tan\beta, \tan\beta)$, and μ_s approaches the horizontal asymptote $\mu^0 = \frac{\eta}{\sin\beta\,\cos\beta} + \tan\beta$ as $\tan\theta$ tends to infinity.

7.2.3 Case 3: sliding down and no tumbling

In this case $\alpha = 0$ and $f = \mu N$. Equations (7.2) may be solved for the three unknowns N, a, and ξ. One finds,

$$N = mg\cos\theta, \tag{7.16}$$

$$a = g\cos\theta\,(\tan\theta - \mu), \tag{7.17}$$

$$\xi = \frac{h}{2}\mu. \tag{7.18}$$

The conditions $a \geq 0$ and $\xi \leq b/2$ lead to

$$\tan\theta \geq \mu$$

$$\mu \leq \tan\beta, \tag{7.19}$$

respectively. Notice that, for these small values of μ ($\mu \leq \tan\beta$), we can make θ as large as we want and the box will never tumble (though it may slide, or not), independently of the exact value of β. This illustrates the crucial role played by friction in coupling the linear and rotational modes. For small values of μ, friction is too small to play this role efficiently.

7.2.4 Case 4: sliding down and tumbling forward

In this case $\xi = b/2$ and $f = \mu N$. Equations (7.2) may be solved for the three unknowns N, a, and α. One finds,

$$\frac{N}{mg\cos\theta} = \frac{\eta}{\eta + \sin^2\beta - \mu\sin\beta\cos\beta}, \tag{7.20}$$

$$\frac{a}{g\cos\theta} = \tan\theta + \frac{\sin\beta\cos\beta - \mu\left(\eta + \cos^2\beta\right)}{\eta + \sin^2\beta - \mu\sin\beta\cos\beta} \tag{7.21}$$

$$\alpha = \frac{g\cos\theta}{R\sin\beta}\left[\frac{\mu - \tan\beta}{\frac{\eta}{\sin\beta\cos\beta} + \tan\beta - \mu}\right]. \tag{7.22}$$

The conditions $N \geq 0$ and $\alpha \geq 0$ lead to

$$\begin{aligned} \mu &\leq \mu^0 = \frac{\eta}{\sin\beta\cos\beta} + \tan\beta, \\ \tan\beta \leq \mu &\leq \mu^0, \end{aligned} \tag{7.23}$$

respectively. Therefore, although here the condition $N \geq 0$ is not trivially satisfied, it is contained in the condition $\alpha \geq 0$. In addition, the condition $a \geq 0$ leads to

$$\mu \leq \frac{\left(\eta + \sin^2\beta\right)\tan\theta + \sin\beta\cos\beta}{\eta + \cos^2\beta + \sin\beta\cos\beta\tan\theta}. \tag{7.24}$$

7.2.5 Summary

The result of the analysis for the case in which the box starts out resting evenly on the plane is presented in Fig. 7.2. The coordinate μ refers to μ_s when there is no sliding (cases 1 and 2), and it refers to our loose symbol μ when there is sliding (cases 3 and 4). We stress the following features:

- In cases 1 and 3 there is no tumbling because of the judicious positioning of the line of action of the normal force: $\xi = h\tan\theta/2$ when there is no sliding; $\xi = h\mu/2$ when the box slides down the plane. When ξ reaches $b/2$ we hit the tumbling regime.

- The line of action of the normal force only passes through G when $\tan\theta = 0$ (i.e., when the box is placed on an horizontal plane), or when $\mu = 0$ (i.e., when the box slides down a frictionless plane).

- It is now easy to understand why spheres (but not boxes) must necessarily roll down inclined planes in the presence of friction. This behavior is rooted in the position of the normal force. In the case of the sphere, there is only one contact point between the object and the inclined plane; and the line of action of the normal force must pass through the center of mass. As

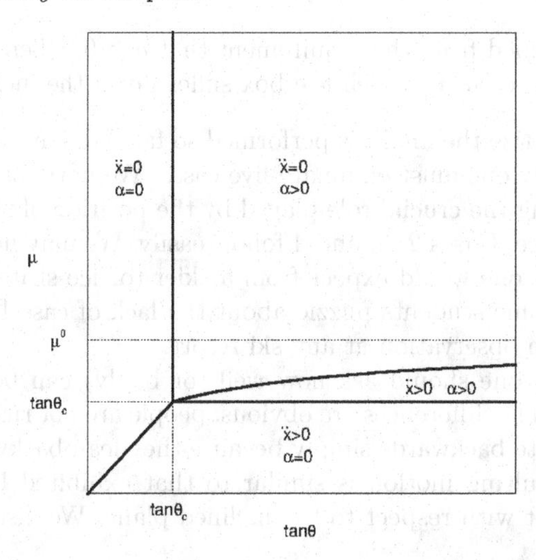

FIGURE 7.2
Structure of $(\tan\theta, \mu)$ parameter space, describing the types of motion allowed for a box resting evenly on an inclined plane. a is the linear acceleration of the lower edge of the box and α is the angular acceleration of the box. The coordinate μ stands for μ_s when $a = 0$, and μ^0 is defined in Eq. (7.23).

a result, there is a net torque with respect to G, which is due to friction and which forces the sphere to roll down the plane. This is only absent for frictionless surfaces. On the contrary, a box has many contact points with a plane when it rests evenly on it. As a result, there are many candidate points for positioning the normal force, and Nature, if it can (as in cases 1 and 3), chooses that point which precludes rotation. For example, in the case with no sliding (case 1), this point is given by $\xi = h\tan\theta/2$. Geometrically, this is the point where the lines of action of the friction and the weight intercept (as we would have expected). The box will only tumble if this point lies outside of the base of the box, since we can obviously not draw the normal force there.

- The straight lines ($\mu = \tan\beta$ and $\tan\theta = \tan\beta$) separating the tumbling and no tumbling regimes arise from the requirement that $\xi \leq b/2$, when one approaches them from the no tumbling cases, while they arise from the requirement that $\alpha \geq 0$, when one approaches them from the tumbling cases.

- It is also interesting to consider the lines separating the sliding—Eq. (7.9) and Eqs. (7.13) and (7.14)—and no sliding cases—Eq. (7.19) and Eqs. (7.23) and (7.24). These lines are obtained from the requirement that $f_s \leq \mu_s N$, when one approaches them from the no sliding cases, and they are obtained from the requirement that $a \geq 0$, when one approaches them from the cases in which the box slides down the inclined plane.

Let us summarize the analysis performed so far. We have analized Table 1 and explained why one must eliminate five cases. We have discussed case 1 in detail, highlighting the crucial role played by the position of the line of action of the normal force. Cases 2, 3, and 4 follow easily. We may now compare our results with what one would expect from a skier (or ice skater) placed on an inclined plane. Some students puzzle about the lack of case DB, since this is the most common observation at any ski resort.

At this point, one should ask how well (or badly) can boxes be used to represent skiers. The differences are obvious: people are not rigid bodies; skiers may fall and rotate backwards simply because they lean backwards too much. However, the resulting motion is similar to that exhibited by boxes placed with an initial tilt with respect to the inclined plane. We discuss this case in the next section.

7.3 Boxes tilted with respect to the plane

Let us now consider the situation depicted in Fig. 7.3, in which the box is positioned initially at rest, but tilted counter-clockwise, making an angle φ with the inclined plane. There are three differences between this case and the

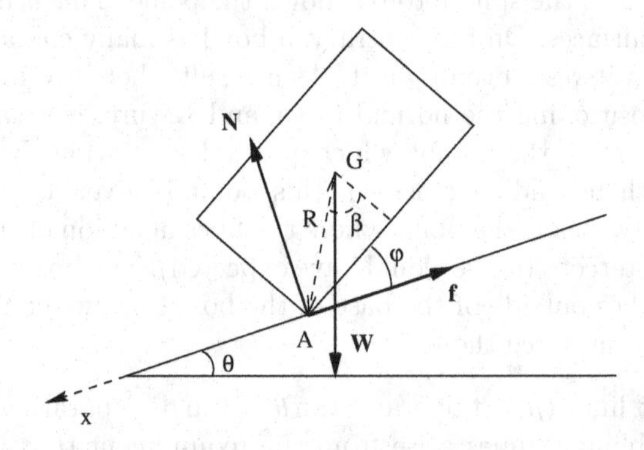

FIGURE 7.3
Box tilted with respect to the plane, but "not excessively": $0 < \varphi \leq \beta$. This corresponds to $\sin \bar{\beta} \geq 0$, where $\bar{\beta} = \beta - \varphi$.

one studied in the previous section. Firstly, the normal force is forced onto the lower edge of the box ($\xi = b/2$), which is now the only point of contact (line of contact, if we do not neglect the dimension perpendicular to the plane of

Fig. 7.3) between the box and the plane. Secondly, the relation between the acceleration of the center of mass (a_G) and the angular acceleration ($\alpha = \ddot{\varphi}$) is altered by the presence of a nonzero φ according to

$$\begin{aligned} (a_G)_x &= a + \alpha R \cos(\beta - \varphi), \\ (a_G)_y &= \alpha R \sin(\beta - \varphi). \end{aligned} \tag{7.25}$$

Thirdly, the torques of the friction and the normal forces, with respect to G, are given by

$$\begin{aligned} &f R \cos(\beta - \varphi), \\ -&N R \sin(\beta - \varphi). \end{aligned} \tag{7.26}$$

As a result, the dynamical equations for $t = 0$ are now given by

$$\begin{cases} N - mg\cos\theta &= m\alpha R \sin\bar{\beta} \\ mg\sin\theta - f &= ma + m\alpha R\cos\bar{\beta}, \\ fR\cos\bar{\beta} - NR\sin\bar{\beta} &= m\eta R^2 \alpha \end{cases} \tag{7.27}$$

where we have defined

$$\bar{\beta} = \beta - \varphi. \tag{7.28}$$

These equations look exactly like those in the cases 2 and 4 described above with the substitution $\beta \to \bar{\beta}$, because there the imminent tumbling also forced $\xi = b/2$. But, here, no other values for ξ are possible and we will not find the continuous range of solutions described in cases 1 and 3 for which $\alpha = 0$. This is the first important difference that this case exhibits, when compared with the previous analysis. The other is more subtle: while β can only take values between 0 and $+\pi/2$, $\bar{\beta}$ may be negative, ranging between $-\pi/2$ and $+\pi/2$. It is precisely this fact that allows solutions in which the lower edge of the box slides up the plane while the box tumbles forward. We will study first the $0 < \varphi \leq \beta$ case (where $\sin\bar{\beta} \geq 0$), leaving the $\beta < \varphi < \pi/2$ case (where $\sin\bar{\beta} < 0$) for last.

Equations (7.27) were obtained for $a \geq 0$ and f positive. As explained in Sec. 7.2, since we start from rest, the sign of $\dot{x}$ after the initial instant is that of a, and our discussion of the $a > 0$ cases applies also for $\dot{x}(0) > 0$. We can extend Eqs. (7.27) for $a < 0$ (in fact, for $a < 0$ and $\dot{x}(0) \leq 0$) with $f \to -f$. Of course, this applies both to Eqs. (7.27) and to their solutions.

One might wonder about the case in which the box is positioned initially at rest, but tilted clockwise, making an angle ϕ with the inclined plane. This case can be reduced trivially to that of a counter-clockwise rotation of an angle $\varphi = \pi/2 - \phi$ by interchanging the height and the length of the box, that is, by transforming simultaneously $\beta \to \pi/2 - \beta$. As a result, the analysis is exactly equal to that which we are about to make, with the trivial substitution $\bar{\beta} \to -\bar{\beta}$.

7.3.1 The case where $0 < \varphi \leq \beta$

The situation depicted in Fig. 7.3 illustrates this case, for which $\sin\bar{\beta} \geq 0$. The crucial point is that the line AG joining the lower edge with the center of mass lies "upward" from the normal to the inclined plane (we could say that the object is leaning forward, but *not excessively*).

Clearly, the analysis for $\alpha \geq 0$ is exactly the same as before (cases 2 and 4 above). In particular, the solutions for $a = 0, \alpha \geq 0$ are still given by Eqs. (7.10) through (7.12), and their limits of validity by Eq. (7.14), only altered by the trivial substitution $\beta \to \bar{\beta}$. Similarly, the solutions for $a > 0, \alpha \geq 0$ are still given by Eqs. (7.20) through (7.22), and their limits of validity by Eqs. (7.23) and (7.24), with $\beta \to \bar{\beta}$.

The big difference arises in the cases with $\alpha = 0$. Here, these correspond to lines of zero measure in parameter space. This situation, which is due to the fact that ξ is fixed at $b/2$, means that the solutions found in Eqs. (7.10) through (7.12) for $a = 0$ do not loose their validity when the condition coming from $\alpha = 0$ (namely, $\tan\theta = \tan\bar{\beta}$) is met. Eqs. (7.10) through (7.12), continue to be valid for $\tan\theta \leq \tan\bar{\beta}$, thus permitting solutions in which the object does not slide but tumbles backward.

Similarly, the solutions found in Eqs. (7.20) through (7.22) for $a > 0$ do not loose their validity when the condition coming from $\alpha = 0$ (namely, $\mu = \tan\bar{\beta}$) is met. Eqs. (7.20) through (7.22), continue to be valid for $\mu \leq \tan\bar{\beta}$, thus permitting solutions in which the object slides down while tumbling backward (case DB).

However, the conditions in Eq. (7.14), for $a = 0$, and Eq. (7.24), for $a > 0$, remain valid. These lines separate the regime of no sliding from the regime of sliding down. They remain valid *even when* $\alpha < 0$. When approached from the $a = 0$ regime, these lines arise from the requirement that $f_s \leq \mu_s N$; when approached from the $a > 0$ regime, they arise from the requirement that $a \geq 0$. As a result, the parameter space for $\varphi > 0$ and $\bar{\beta} \geq 0$ presented in Fig. 7.4, differs from that presented in Fig. 7.2.

In Fig. 7.2, there was a straight line joining the point $(\tan\theta, \mu) = (0, 0)$ with the point $(\tan\theta, \mu) = (\tan\beta, \tan\beta)$, which separated the two no tumble regimes (no sliding and sliding down). In Fig. 7.4, the no tumble regimes are represented by lines of measure zero, consistent with the fact that these are unstable configurations (the reader is invited to place a box on edge on an inclined plane in equilibrium—without blowing). Moreover, in Fig. 7.4, the regimes in which the box tumbles backward are separated by the hyperbola in Eq. (7.24), which, as before, passes through $(\tan\theta, \mu) = (\tan\bar{\beta}, \tan\bar{\beta})$ and tends to the horizontal asymptote μ_2^+. However, in this case, the hyperbola starts at $(\tan\theta, \mu) = (0, \mu_1^+)$. Here and in Fig. 7.4, we have used

$$\begin{aligned} \mu_1^+ &= \left[\frac{\eta}{\sin\bar{\beta}\,\cos\bar{\beta}} + \cot\bar{\beta}\right]^{-1}, \\ \mu_2^+ &= \frac{\eta}{\sin\bar{\beta}\,\cos\bar{\beta}} + \tan\bar{\beta}. \end{aligned} \tag{7.29}$$

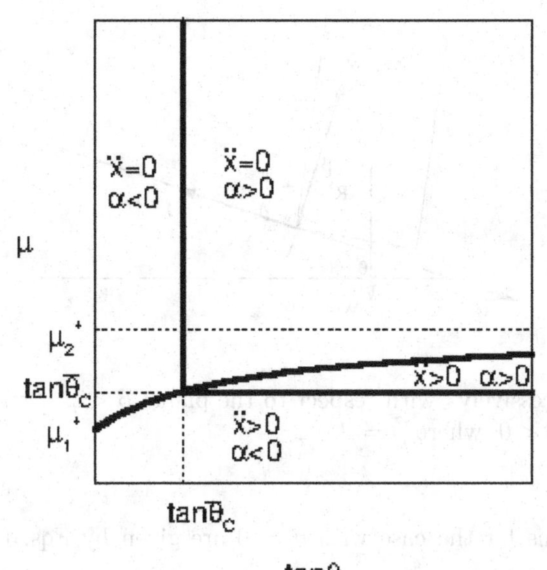

FIGURE 7.4
Structure of $(\tan\theta, \mu)$ parameter space, describing the types of motion allowed for a box tilted with respect to the plane, but "not excessively": $0 < \varphi \leq \beta$. a is the linear acceleration of the lower edge of the box and α is the angular acceleration of the box. The coordinate μ stands for μ_s when $a = 0$. The quantities μ_1^+ and μ_2^+ are defined in Eq. (7.29).

Clearly, $0 < \mu_1^+ < \tan\bar{\beta} < \mu_2^+$. As a result, for small values of μ_s ($\mu_s < \mu_1^+$), the box will necessarily slide, no matter how small the value of $\tan\theta$. In particular, for small enough values of μ, if one tips a bookcase, it will tumble in one direction but, at the same time, it will slide in the opposite direction.

Finally, it is easy to see that no solutions with $\bar{\beta} > 0$ exist in which the lower edge of the box slides up the plane.

7.3.2 The case where $\beta < \varphi < \pi/2$

The situation depicted in Fig. 7.5 falls under this case, for which $\sin\bar{\beta} < 0$. The crucial point is that the line AG joining the lower edge with the center of mass lies "downward" from the normal to the inclined plane (we could say that the object is leaning forward *excessively*).

Again, the Eq.(7.27) involved in this case are similar to those involved in case 2 (for $a = 0$) and in case 4 (for $a > 0$), with the substitution $\beta \to \bar{\beta}$. In particular,

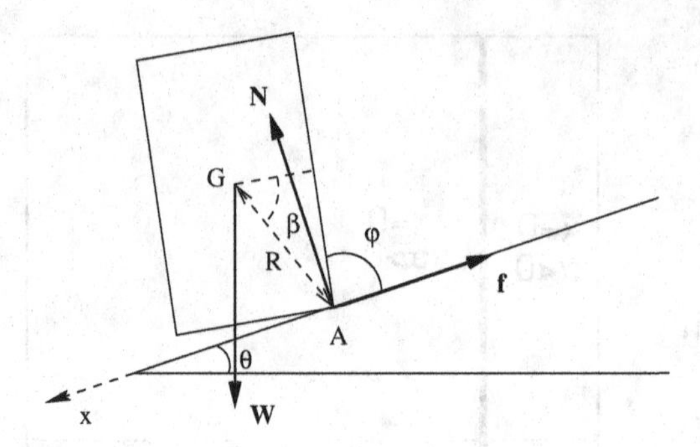

FIGURE 7.5
Box tilted "excessively" with respect to the plane: $\beta < \varphi < \pi/2$. This corresponds to $\sin\bar{\beta} < 0$, where $\bar{\beta} = \beta - \varphi$.

- the solutions for the case with $a = 0$ are given by Eqs. (7.10) through (7.12);
- the solutions for the case with $a > 0$ are given by Eqs. (7.20) through (7.22);
- the solutions for the case with $a < 0$ coincide with those found in the case $a > 0$, namely Eqs. (7.20) through (7.22), with the substitution $f \to -f$.

However, since now $\sin\bar{\beta} \leq 0$, the validity conditions are very different, leading to a completely different structure in parameter space.

For later use, it will be convenient to define the quantities

$$\begin{aligned} \mu_1^- &= -\mu_1^+ , \\ \mu_2^- &= -\mu_2^+ , \\ \tan\theta_1 &= 1/\mu_2^- , \\ \tan\theta_2 &= 1/\mu_1^- . \end{aligned} \tag{7.30}$$

Clearly,

$$0 \leq \mu_1^- \leq -\tan\bar{\beta} \leq \mu_2^- , \tag{7.31}$$

and

$$0 \leq \tan\theta_1 \leq -\cot\bar{\beta} \leq \tan\theta_2 . \tag{7.32}$$

7.3.2.1 The case of $\beta < \varphi < \pi/2$ and $a = 0$

The solutions in this case coincide with those found in Eqs. (7.10) through (7.12), with $\beta \to \bar{\beta}$. However, since $\sin\bar{\beta} < 0$, the conditions $N \geq 0$ and $f_s \geq 0$ are not trivially satisfied.

Since $\sin\bar{\beta} < 0$, the condition $N \geq 0$ leads to

$$\tan\theta \leq \tan\theta_2. \tag{7.33}$$

It is easy to see from Eq. (7.11) that f_s may be positive or negative, in contrast with what happened in the case with $\sin\bar{\beta} \geq 0$, in which f_s could only take positive values. In fact, as we shall now see, the friction may be negative (pointing downwards along the plane) because in its absence the edge would slip up the plane for small values of $\tan\theta$. As a result, we should look for the physical condition $|f_s| \leq \mu_s N$, leading to

$$\mu_s \geq \frac{\left|(\eta + \sin^2\bar{\beta})\tan\theta + \sin\bar{\beta}\cos\bar{\beta}\right|}{-\eta - \cos^2\bar{\beta} - \sin\bar{\beta}\cos\bar{\beta}\tan\theta} . \tag{7.34}$$

Notice that the denominator (with the explicit minus sign) is related with N and positive. The presence of the magnitude on the right hand side of this equations generates two different regimes. In the first regime f_s is negative and

$$\begin{aligned} &\tan\theta < \tan\theta_1, \\ &\mu_s \geq \frac{(\eta + \sin^2\bar{\beta})\tan\theta + \sin\bar{\beta}\cos\bar{\beta}}{-\eta - \cos^2\bar{\beta} - \sin\bar{\beta}\cos\bar{\beta}\tan\theta} . \end{aligned} \tag{7.35}$$

After dividing the numerator and the denominator by $\sin\bar{\beta}\cos\bar{\beta}$ and using Eqs. (7.29) and (7.30), the above inequality may be rewritten in a more compact form

$$\mu_s \geq \frac{\tan\theta_1 - \tan\theta}{\tan\theta_2 - \tan\theta}\mu_2^- . \tag{7.36}$$

This defines a curve (again a branch of hyperbola) which starts out at $\mu_s = \mu_1^-$ for $\tan\theta = 0$ and drops to $\mu_s = 0$ for $\tan\theta = \tan\theta_1$. The second regime of validity for Eq. (7.34) corresponds to f_s positive and is defined by

$$\begin{aligned} &\tan\theta_1 \leq \tan\theta < \tan\theta_2, \\ &\mu_s \geq \frac{(\eta + \sin^2\bar{\beta})\tan\theta + \sin\bar{\beta}\cos\bar{\beta}}{\eta + \cos^2\bar{\beta} + \sin\bar{\beta}\cos\bar{\beta}\tan\theta} . \end{aligned} \tag{7.37}$$

The right hand side of the last equation coincides with that of Eq. (7.35), except for the overall minus sign. Thus, it may be rewritten as

$$\mu_s \geq \frac{\tan\theta_1 - \tan\theta}{\tan\theta_2 - \tan\theta}\mu_2^+ . \tag{7.38}$$

This defines another branch of hyperbola which starts out at $\mu_s = 0$ for $\tan\theta = \tan\theta_1$ and tends to infinity as $\tan\theta$ approaches $\tan\theta_2$.

Finally, since $\sin\bar{\beta} \leq 0$, Eq. (7.12) guarantees that $\alpha \geq 0$ for all the values of parameter space.

7.3.2.2 The case of $\beta < \varphi < \pi/2$ and $a > 0$

The solutions in this case coincide with those found in Eqs. (7.20) through (7.22), with $\beta \to \bar{\beta}$. Here, the condition $N \geq 0$ is trivially satisfied. On the other hand, the condition $a \geq 0$ is always satisfied for $\tan\theta \geq \tan\theta_2$, but it provides the nontrivial constraint

$$\mu \leq \frac{(\eta + \sin^2\bar{\beta})\tan\theta + \sin\bar{\beta}\cos\bar{\beta}}{\eta + \cos^2\bar{\beta} + \sin\bar{\beta}\cos\bar{\beta}\tan\theta}, \tag{7.39}$$

for $\tan\theta < \tan\theta_2$. The right hand side of this equation coincides with the right hand side of Eq. (7.37), and, thus, we have the alternative form for this inequality

$$\mu_s \leq \frac{\tan\theta_1 - \tan\theta}{\tan\theta_2 - \tan\theta}\mu_2^+ . \tag{7.40}$$

When we join our ambiguous μ and μ_s into one figure, this implies that the regime with $a > 0$ lies to the right of the regime with $a = 0$ in parameter space.

Again, since $\sin\bar{\beta} \leq 0$, Eq. (7.22) guarantees that $\alpha \geq 0$ for all the values of parameter space.

7.3.2.3 The case of $\beta < \varphi < \pi/2$ and $a < 0$

The solutions in this case coincide with those found in Eqs. (7.20) through (7.22), with $\beta \to \bar{\beta}$ and the change $\mu \to -\mu$, which is due to the fact that the friction points downwards along the plane when the edge of the box slides up the plane.

The condition $N \geq 0$ implies that

$$\mu \leq \mu_2^- . \tag{7.41}$$

The requirement that $a < 0$ must be imposed on Eq. (7.21) after the changes mentioned above. Dividing out again the factor $\sin\bar{\beta}\cos\bar{\beta}$, and using Eqs. (7.29) and (7.30), we obtain

$$\frac{\tan\theta\cot\theta_1 - 1 + \mu(\tan\theta_2 - \tan\theta)}{\mu_2^- - \mu} < 0. \tag{7.42}$$

Using the constraint (7.41) due to $N \geq 0$, this becomes

$$\mu(\tan\theta_2 - \tan\theta) < 1 - \tan\theta\cot\theta_1. \tag{7.43}$$

Since $\theta_1 < \theta_2$, we could have either $\theta < \theta_1 < \theta_2$ or $\theta_1 < \theta_2 < \theta$. The second alternative is incompatible with Eq. (7.41) since it would imply that

$$\mu > \frac{\tan\theta - \tan\theta_1}{\tan\theta - \tan\theta_2}\cot\theta_1 > \mu_2^- . \tag{7.44}$$

Therefore, we are lead to the inequalities

$$\tan\theta < \tan\theta_1,$$
$$\mu < \frac{\tan\theta - \tan\theta_1}{\tan\theta - \tan\theta_2}\mu_2^- . \tag{7.45}$$

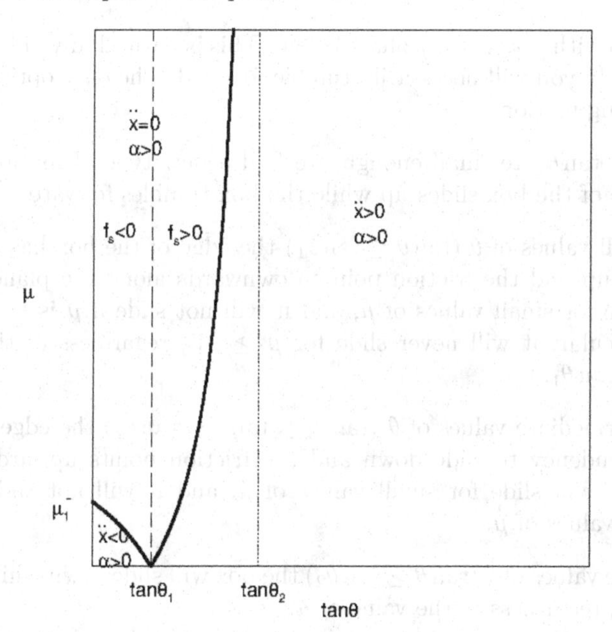

FIGURE 7.6
Structure of $(\tan\theta, \mu)$ parameter space, describing the types of motion allowed for a box tilted "excessively" with respect to the plane: $\beta < \varphi < \pi/2$. a is the linear acceleration of the lower edge of the box and α is the angular acceleration of the box. The coordinate μ stands for μ_s when $a = 0$. The quantities μ_1^-, $\tan\theta_1$, and $\tan\theta_2$ are defined in Eq. (7.30). For all these cases $\alpha > 0$.

Comparing this equation with Eq. (7.36), we conclude that the solutions with $a < 0$ lie below the solutions with $a = 0$ and negative friction. In fact, the solutions found here lie below the curve which starts out at $\mu = \mu_1^-$ for $\tan\theta = 0$ and drops to $\mu = 0$ for $\tan\theta = \tan\theta_1$.

Finally, the combination of the $N \geq 0$ and $a \leq 0$ requirements, leads to $\alpha \geq 0$ in all of the region of parameter space allowed by those conditions.

7.3.2.4 Summary

The results which we have found for $\beta < \varphi < \pi/2$ ($\sin\bar{\beta} < 0$) are summarized in Fig. 7.6. The following features of Fig. 7.6 are worth noticing:

- The lines separating the $a = 0$ and $a \neq 0$ regions arise from the requirement that $|f_s| \leq \mu_s N$, when they are approached from the $a = 0$ region, and from the requirement that $a > 0$ ($a < 0$), when they are approached from the sliding down (sliding up) region.

- All cases with $\beta < \varphi < \pi/2$ have $\alpha \geq 0$. This is natural: if you lean forward excessively you will necessarily tumble forward. The only options concern the sliding motion.

- If μ and $\tan\theta$ are small enough, we find a new type of motion in which the edge of the box slides up while the box tumbles forward.

- For small values of θ ($\tan\theta \leq \tan\theta_1$) the edge of the box has a tendency to slide up and the friction points downwards along the plane. The box will slide for small values of μ, and it will not slide if μ is large enough. In particular, it will never slide for $\mu \geq \mu_1^-$, regardless of the value of $\tan\theta \leq \tan\theta_1$.

- For intermediate values of θ ($\tan\theta_1 \leq \tan\theta \leq \tan\theta_2$) the edge of the box has a tendency to slide down and the friction points upward along the plane. It will slide for small values of μ, and it will not slide for large enough values of μ.

- For large values of θ ($\tan\theta \geq \tan\theta_2$) the box will slide down while tumbling forward, regardless of the value of μ.

- The region of parameter space where $a < 0$ seems unduly small. This is due to the fact that we have chosen $\tan\theta$ as our coordinate axis. This region would be much expanded if θ was chosen as the coordinate axis. Nevertheless, this region is bounded by $\tan\theta_1$, which reaches its maximum value of 3/4 when $\varphi - \beta = \arctan 1/2$. As a result, for θ greater than $\arctan 3/4 \approx 36.9^o$, there is no combination of box shape (β) and initial tilt (φ) that will make the edge of the box slide up the plane. The friction will always point up; any given box will either not slide at all (and this only up to $\tan\theta = \tan\theta_2$) or slide down the slope. The same result holds for μ greater than 3/4, since this is the maximum value of μ_1^- as a function of $\varphi - \beta$.

7.4 Conclusions

We have studied the problem of a box placed initially at rest on an inclined plane, performing a detailed analysis of the various regimes of motion. A quick glance at Figs. 7.2, 7.4, and 7.6 shows how rich the structure of this problem is. As a result, this problem can be used to understand (and gain intuition about) a very large range of everyday situations in which friction plays an important role.

Remarkably, most textbooks contain only a brief reference to this problem; and this only for some particular cases. We believe that students may gain a great deal (knowledge, intuition, and fun) by being exposed to Figs. 7.2, 7.4, and 7.6 in lectures on the connection between friction and rotation. In particular, the possibility that objects may fall backwards while slipping down the plane (a typical ski slope event included in Fig. 7.4), and the possibility that objects may tumble forwards while slipping up the plane (a classic cartoon situation included in Fig. 7.6), are bound to captivate the students.

On a more quantitative aspect, the techniques used in finding the limits of validity of the various types of motion involve most of the relevant features of friction, thus highlighting the tools needed for a large variety of other problems.

Acknowledgments

We thank A. Barroso and A. J. Silvestre for discussions. We are also indebted to B. Braizinha for help with the software used in drawing the figures.

References

[1] D. Halliday, R. Resnick, and J. Walker, *Fundamentals of Physics–Extended*, 5th ed. (John Wiley and Sons, New York, 1997).

[2] P. M. Fishbane, S. Gasiorowicz, and S. T. Thornton, *Physics for Scientists and Engineers*, 2nd ed. (Prentice Hall, NJ, 1996).

[3] T. L. Chow, *Classical Mechanics* (John Wiley and Sons, New York, 1995).

[4] L. N. Hand and J. D. Finch, *Analytical Mechanics* (Cambridge University Press, Cambridge, 1998).

8

Magnetic Forces Acting on Rigid Current-Carrying Wires Placed in a Uniform Magnetic Field

A. Casaca and J. P. Silva

CONTENTS

We calculate the forces acting on segments of zero thickness rigid wires carrying constant currents and placed in a uniform magnetic field. This example entices the students to formulate, explore, and prove a conjecture, exposing them to an early example of a research-like project. The sequence of examples discussed here was literally born during an introductory physics class period dedicated to the calculation of the forces acting on segments of current-carrying wires placed in a uniform magnetic field. On that occasion, a series of interchanges between the lecturer and the actively participating students led to the formulation of a conjecture, its exploration, and its final resolution. The resulting lecture was then spontaneously turned into an introduction (of sorts) to research in physics.

Consider the rigid wires shown in Figs. 8.1(a) and 8.1(b), which carry the constant current I and are placed in a region with uniform magnetic field $\vec{B}$, perpendicular to the page and directed inwardly. The resulting magnetic forces acting on these wires are easily calculated, yielding [1]

$$\vec{F} = \int I d\vec{l} \times \vec{B} = I \left(\int_0^{2R} dl \right) B(-\mathbf{j}) = -I(2R)B\mathbf{j}, \tag{8.1a}$$

Reproduced from A. Casaca and J. P. Silva, "Magnetic Forces Acting on Rigid Current-Carrying Wires Placed in a Uniform Magnetic Field," The Physics Teacher **42**, 161–163 (2004), https://doi.org/10.1119/1.1664383, with the permission of the American Association of Physics Teachers.

DOI: 10.1201/9781003187103-8

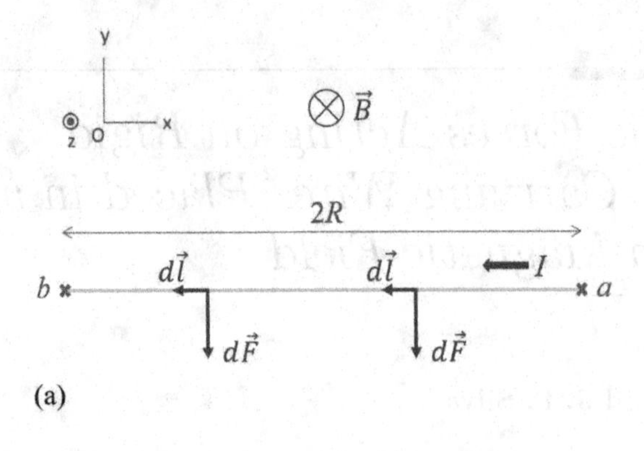

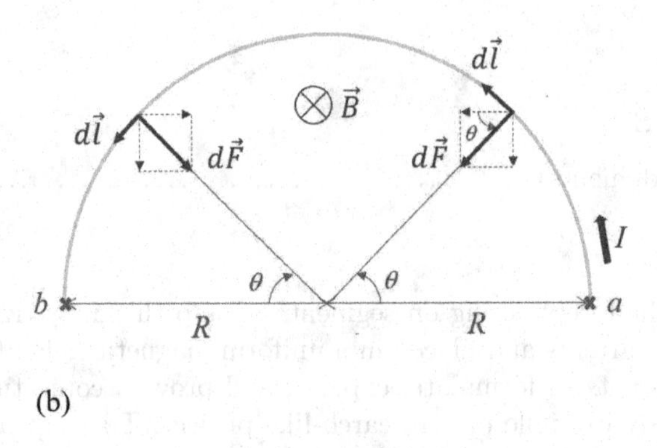

FIGURE 8.1
Rigid wires, carrying a constant current I, and placed in a region of uniform magnetic field $\vec{B}$. (a) Straight-line segment of wire with length $2R$. (b) Semicircular segment of wire with radius R.

and

$$\vec{F} = \int I dl\, B \sin\theta\,(-\mathbf{j}) = IB \int_0^{2\pi} \sin\theta\, R d\theta(-\mathbf{j}) = -I(2R)B\mathbf{j}, \qquad (8.1b)$$

respectively. Here, $\mathbf{j}$ is the unit vector pointing vertically upward.

The result is the same in both cases! At first sight, this may be surprising. On the one hand, all the infinitesimal forces acting on the length elements of the straight-line segment of wire point vertically downward. On the other hand, the forces acting on each length element of the semicircular segment of wire point toward the center of the circle. It seems to be the symmetry of the semicircle that guarantees the cancellation of the horizontal components in the sum, leading to an overall resultant force that points vertically downward.

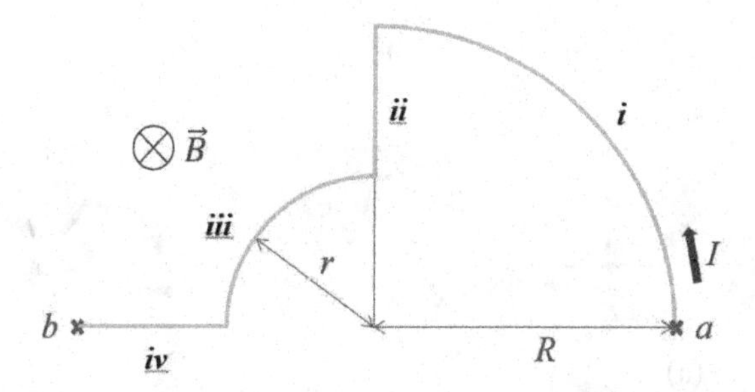

FIGURE 8.2
Segment of current-carrying wire built out of straight lines and of quarter-circle segments, and placed in a region of uniform magnetic field $\vec{B}$.

Is the equality between Eqs. (8.1a) and (8.1b) a coincidence, or does the result hold for any shape?

Our conjecture is that this result holds in general. We test it further by considering the asymmetric wire in Fig. 8.2. The result can be found by looking at each segment separately. We will need the force acting on a wire with the shape of a quarter-circle. This can be calculated in a creative way by noticing that two quarter-circles make up a semicircle and that, by symmetry, the force acting on a quarter-circle must point along an angle of 45°. Combining these observations with Eq. (8.1b), we conclude that the quarter-circle of radius R (i) is subject to a force given by

$$\vec{F}_{\text{i}} = -IRB\mathbf{i} - IRB\mathbf{j}, \tag{8.2a}$$

where the unit vector points horizontally to the right. Continuing from right to left, the forces on the other segments are easily found to be

$$\vec{F}_{\text{ii}} = I(R-r)B\mathbf{i}, \tag{8.2b}$$

$$\vec{F}_{\text{iii}} = IrB\mathbf{i} - IrB\mathbf{j}, \tag{8.2c}$$

$$\vec{F}_{\text{iv}} = -I(R-r)B\mathbf{j}. \tag{8.2d}$$

The resulting force acting on the wire shown in Fig. 8.2 is given by

$$\vec{F} = \vec{F}_{\text{i}} + \vec{F}_{\text{ii}} + \vec{F}_{\text{iii}} + \vec{F}_{\text{iv}} = -I(2R)B\mathbf{j}, \tag{8.2e}$$

which coincides with the results in Eqs. (8.1a) and (8.1b). It is interesting that this result arises from a conspiring cancellation of the various horizontal components, even though Fig. 8.2 is not left-right symmetric.

We are now fairly confident in our conjecture, and we should seek to prove it for a wire of arbitrary shape. Our objective is to prove that the force acting on the two wires shown in Fig. 8.3(a) is the same. This is clearly the same as

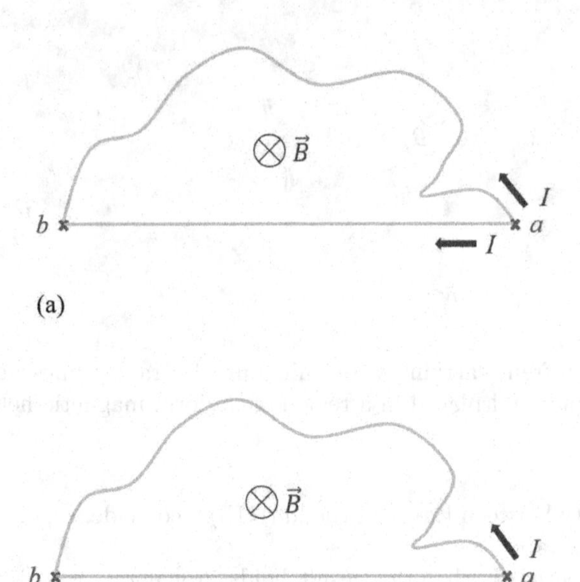

FIGURE 8.3
(a) Comparison between a straight-line wire and a wire of arbitrary shape, both placed in a region of uniform magnetic field $\vec{B}$, and carrying a constant current I between the same two points a and b. (b) Closed, current-carrying circuit obtained from Fig. 8.3(a) by reversing the current on the straight-line wire.

proving that the force acting on the closed circuit in Fig. 8.3(b) is zero [1], since the force reverses its sign when the current I reverses its sign. But this is a simple consequence of our assumption that $\vec{B}$ is a uniform field. Indeed, under this assumption, the force acting on an arbitrary, closed, rigid current-carrying wire is [2]

$$\vec{F} = \oint I d\vec{l} \times \vec{B} = I \left(\oint d\vec{l} \right) \times \vec{B} = \vec{0}, \tag{8.3}$$

because for a closed circuit, the integral of all the elementary vectors along the path vanishes (as it had to since we come back to the same point, thus incurring no displacement). In addition, we learn that need not be

[1] A closed circuit of arbitrary shape carrying a constant current I has a magnetic moment $\vec{m}$; when the circuit is placed in a uniform magnetic field $\vec{B}$, a torque will result on the magnetic moment given by $\vec{\tau} = \vec{m} \times \vec{B}$, which tends to align the magnetic moment with the external magnetic field.

perpendicular to the plane of the circuit and that, in fact, the circuit need not even be planar.

Therefore, our initial conjecture is reinforced: Any rigid wire connecting two points, a and b, of any length or shape, planar or not, carrying a constant current, experiences the same magnetic force if placed in a uniform magnetic field.

Our students reached another interesting conclusion: The professors are "mean people" because they force the students to calculate the forces acting on wires with the complicated geometries of Figs. 8.1(b) and 8.2,even though they know that the result for a straight wire is much simpler and leads to exactly the same result.

It is interesting that such a simple example of the three fundamental steps of any research project (namely the conjecture that springs from some calculation/observation, the consideration of simple examples to build confidence in the result, and the search for a general proof) has arisen naturally from the questions of students in an otherwise rather ordinary recitation of elementary physics. We and our students had great fun. We hope others will too.

Acknowledgments

The authors would like to thank their students for their excitement when this problem "exploded" in class, and for their suggestion that we share it with others. We are indebted to A. J. Silvestre for reading this manuscript.

References

[1] These results may be found in any introductory physics manual. See, for example, [2].

[2] This result is proven, for example, in R. A. Serway, *Physics for Scientists and Engineers with Modern Physics*, 3rd ed. extended version (Saunders, 1992).

9

Comparing a Current-Carrying Circular Wire with Polygons of Equal Perimeter: Magnetic Field versus Magnetic Flux

J. P. Silva and A. J. Silvestre

CONTENTS

We compare the magnetic field at the center of and the self-magnetic flux through a current-carrying circular loop, with those obtained for current-carrying polygons with the same perimeter. As the magnetic field diverges at the position of the wires, we compare the self-fluxes utilizing several regularization procedures. The calculation is best performed utilizing the vector potential, thus highlighting its usefulness in practical applications. Our analysis answers some of the intuition challenges students face when they encounter a related simple textbook example. These results can be applied directly to the determination of mutual inductances in a variety of situations.

DOI: 10.1201/9781003187103-9

9.1 Introduction

A common exercise in introductory physics courses concerns the comparison between the magnetic fields due to two loops of equal length P, carrying the same current i, one shaped into a square and the other shaped into a circle. One is asked to compare the magnetic fields at the centers of the respective figures [1], finding that the field at the center of the square is larger than the field at the center of the circle. In our classes, this problem is always followed by a lively debate. Many students feel that the opposite should occur, citing the fact that, for a given perimeter P, the circle is the figure with the largest area. It is only when the two figures are drawn to scale, as in Fig. 9.1, that

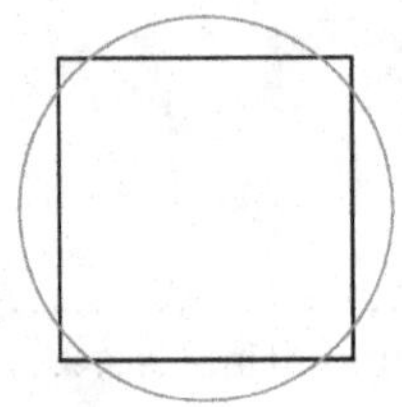

FIGURE 9.1
Square and circle of equal perimeter P.

they understand the result. The point is that, for equal perimeter, the sides of the square lie inside the circle for most of the integration paths.

The result can be easily generalized for any polygon with n equal sides and total perimeter P. Figure 9.2 illustrates the case of $n = 5$. Each side has length $s_n = P/n$, placed at a distance $d_n = s_n/2 \cot(\theta_n/2)$ from the center, where $\theta_n = 2\pi/n$. The total magnetic field is simply equal to n times the field produced by a straight wire of length s_n carrying a current i, at a point placed at a distance d_n from the wire, along its perpendicular bisector:

$$B_n^{\text{center}} = n \frac{\mu_0 i}{4\pi d_n} \frac{s_n}{\sqrt{(s_n/2)^2 + d_n^2}} = \frac{\mu_0 i}{4\pi P} 4n^2 \tan(\pi/n) \sin(\pi/n). \tag{9.1}$$

Substituting for $n = 3, 4, \ldots$ in Eq. (9.1), we conclude that, for equal perimeter, the field at the center of a current-carrying triangle is the largest; and the fields at the center of other current-carrying polygons with equal perimeter decrease as the number of sides increases, approaching the asymptotic value of $B_{\text{c}}^{\text{center}} = \frac{\mu_0 i}{4\pi P} 4\pi^2$ obtained for the circle. This calculation can be assigned as a homework exercise.

Although the area does not play a role in this example, our students usually point out that it should play a role in determining the auto-flux through the wire loops. For a given perimeter P, the areas enclosed by the polygon wires

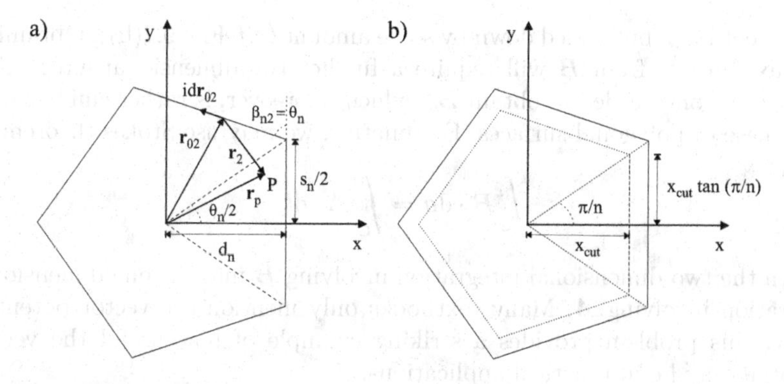

FIGURE 9.2
Pentagon with perimeter P. (a) Pictorial representation of the vectors used in the calculation of $\boldsymbol{A}$, which are defined in the text. (b) The line integral of $\boldsymbol{A}$ is taken along the inner (dotted) polygonal curve C_n.

are $\mathcal{A}_n = P^2 \cot(\pi/n)/(4n)$, approaching the area of the circle, $\mathcal{A}_c = P^2/(4\pi)$, as the number of sides increases. The naive multiplication

$$B_n^{\text{center}} \mathcal{A}_n = \frac{\mu_0 i P}{4\pi} n \sin(\pi/n), \tag{9.2}$$

grows with n. Normalizing this type of "flux" by $\frac{\mu_0 i P}{4\pi}$, as we shall henceforth do[1], we find 2.6, 2.8, 3.1, and π for $n = 3$, $n = 4$, $n = 8$, and the circle, respectively. This seems to indicate that the smaller field at the center of the circle is more than compensated by its larger area. Some students interpret this as a vindication of their initial intuition.

Unfortunately, things are considerably more complicated than this simple argument suggests, making it interesting to revisit this problem in an advanced course on electromagnetism. Firstly, the magnetic field varies from point to point in space. The calculations of these magnetic fields may be found elsewhere for the polygon [2], for the circular loop [3], and for planar wires [4]. Secondly, these fields diverge at the position of the wires, meaning that some regularization must be used. Thirdly, obtaining the flux directly from the magnetic fields requires a two dimensional integration, which becomes particularly difficult in the case of polygons.

In this article, we start by calculating the vector potential $\boldsymbol{A}$ produced by a circular or polygonal loop of perimeter P and carrying a current i, at any point in the plane of the figure, inside the figure. Naturally, $\boldsymbol{A}$ and $\boldsymbol{B} = \nabla \times \boldsymbol{A}$ diverge as one approaches the wire loop. So, we will consider the flux of $\boldsymbol{B}$ through a surface S with edges on a curve C similar to (and concentric with)

[1] All our figures will be drawn for the flux Φ in units of $\mu_0 i P/(4\pi)$, i.e., whenever we mention Φ on the vertical axis, we are really plotting $4\pi\Phi/(\mu_0 i P)$.

the current loop, but scaled down by some amount (*c.f.* Fig. 9.2(b)). Obtaining the flux directly from $\boldsymbol{B}$ will require a further two-dimensional integration (besides the one needed to obtain $\boldsymbol{B}$), which, moreover, is rather cumbersome in the case of polygonal surfaces. Fortunately, we may use Stokes theorem

$$\int_S \boldsymbol{B} \cdot \mathrm{d}\boldsymbol{a} = \int_C \boldsymbol{A} \cdot \mathrm{d}\boldsymbol{l} \tag{9.3}$$

to turn the two-dimensional integration involving $\boldsymbol{B}$ into the one-dimensional integration involving $\boldsymbol{A}$. Many textbooks only mention the vector potential briefly; this problem provides a striking example of how useful the vector potential may be in practical applications.

The results we obtain also provide the mutual inductance of two nested, coplanar, and concentric (polygonal or circular) wires of equal shape but different scales. This can be used for theoretical discussions and experimental studies of Faraday's law.

9.2 Calculating the vector potential

We wish to calculate $\boldsymbol{A}_n(x,y)$ at a point P with coordinates $\boldsymbol{r}_P = x\,\hat{\mathbf{e}}_x + y\,\hat{\mathbf{e}}_y$, as illustrated in Fig. 9.2(a). We start by parametrizing the positions of the points on the right-hand side of the polygon as $\boldsymbol{r}_{01} = d_n\,\hat{\mathbf{e}}_x + t\,\hat{\mathbf{e}}_y$, with $t \in (-s_n/2, s_n/2)$. Using $\boldsymbol{r}_1 = \boldsymbol{r}_P - \boldsymbol{r}_{01}$, we find

$$\begin{aligned} \frac{4\pi}{\mu_0 i}\boldsymbol{A}_{n1} &= \int_{-s_n/2}^{s_n/2} \frac{1}{r_1}\frac{\mathrm{d}\boldsymbol{r}_{01}}{\mathrm{d}t}\mathrm{d}t = \int_{-s_n/2}^{s_n/2} \frac{\mathrm{d}t}{\sqrt{(x-d_n)^2+(y-t)^2}}\,\hat{\mathbf{e}}_y \\ &= \ln\left\{\frac{-y+s_n/2+\sqrt{[x-d_n]^2+[y-s_n/2]^2}}{-y-s_n/2+\sqrt{[x-d_n]^2+[y+s_n/2]^2}}\right\}\hat{\mathbf{e}}_y\,. \end{aligned} \tag{9.4}$$

The position of the points along the k-th side (moving anti-clockwise) is simply given by a rotation of $\boldsymbol{r}_{01}$ by an angle $\beta_{nk} = (k-1)\theta_n = 2\pi(k-1)/n$. So, $\boldsymbol{r}_{0k} = X_{nk}(t)\,\hat{\mathbf{e}}_x + Y_{nk}(t)\,\hat{\mathbf{e}}_y$, where

$$\begin{aligned} X_{nk}(t) &= d_n\cos\beta_{nk} - t\sin\beta_{nk}, \\ Y_{nk}(t) &= d_n\sin\beta_{nk} + t\cos\beta_{nk}. \end{aligned} \tag{9.5}$$

As a result

$$\frac{4\pi}{\mu_0 i}\boldsymbol{A}_{nk} = \int_{-s_n/2}^{s_n/2} \frac{\mathrm{d}t}{\sqrt{[x - X_{nk}(t)]^2 + [y - Y_{nk}(t)]^2}}\,\hat{\mathbf{e}}_{nk}$$

$$= \ln\left\{\frac{s_n/2 - a_{nk}(x,y) + \sqrt{[x - X_{nk}(s_n/2)]^2 + [y - Y_{nk}(s_n/2)]^2}}{-s_n/2 - a_{nk}(x,y) + \sqrt{[x - X_{nk}(-s_n/2)]^2 + [y - Y_{nk}(-s_n/2)]^2}}\right\}\hat{\mathbf{e}}_{nk}, \tag{9.6}$$

where

$$\hat{\mathbf{e}}_{nk} = -\sin\beta_{nk}\,\hat{\mathbf{e}}_x + \cos\beta_{nk}\,\hat{\mathbf{e}}_y \tag{9.7}$$

and

$$\begin{aligned}\pm s_n/2 - a_{nk}(x,y) &= [x - X_{nk}(\pm s_n/2)]\sin\beta_{nk} \\ &\quad - [y - Y_{nk}(\pm s_n/2)]\cos\beta_{nk}.\end{aligned} \tag{9.8}$$

The final magnetic vector potential is given by

$$\boldsymbol{A}_n(x,y) = \sum_{k=1}^{n}\boldsymbol{A}_{nk}(x,y). \tag{9.9}$$

Alternatively, we might obtain Eq. (9.6) from Eq. (9.4) through the vector field rotations discussed by Grivich and Jackson [2]. We could now recover their Eq. (9.9) with $z = 0$ by taking $\boldsymbol{B} = \nabla\times\boldsymbol{A}$ and suitable variable redefinitions[2].

As for the circular loop, we use polar coordinates. By symmetry,

$$\boldsymbol{A}_c(\rho,\theta) = A_c(\rho,\theta)\,\hat{\mathbf{e}}_\theta = A_c(\rho,0)\,\hat{\mathbf{e}}_\theta, \tag{9.10}$$

and we take $\boldsymbol{r}_P = \rho\,\hat{\mathbf{e}}_x$. Parametrizing the positions of the points along the current-carrying circular wire of radius R as $\boldsymbol{r}_0 = R\cos\varphi\,\hat{\mathbf{e}}_x + R\sin\varphi\,\hat{\mathbf{e}}_y$, with $\varphi\in(0,2\pi)$, $\boldsymbol{r} = \boldsymbol{r}_P - \boldsymbol{r}_0$, and we find

$$\frac{4\pi}{\mu_0 i}\boldsymbol{A}_c(\rho,0) = \int_0^{2\pi}\frac{1}{r}\frac{\mathrm{d}\boldsymbol{r}_0}{\mathrm{d}\varphi}\mathrm{d}\varphi = \int_0^{2\pi}\frac{-R\sin\varphi\,\hat{\mathbf{e}}_x + R\cos\varphi\,\hat{\mathbf{e}}_y}{\sqrt{\rho^2 + R^2 - 2\rho R\cos\varphi}}\mathrm{d}\varphi$$

$$= \frac{2}{\rho(\rho+R)}\left[(\rho^2+R^2)\,K\left(\frac{2\sqrt{\rho R}}{\rho+R}\right) - (\rho+R)^2\,E\left(\frac{2\sqrt{\rho R}}{\rho+R}\right)\right]\hat{\mathbf{e}}_y, \tag{9.11}$$

[2]There is a subtlety concerning the fact that, since we have determined $\boldsymbol{A}(x,y,z)$ only for the plane $z = 0$, we cannot perform the derivations with respect to z. However, these do not enter the calculation of $B_z(x,y,0)$ which, by symmetry, is the only non-vanishing component of $\boldsymbol{B}(x,y,z)$ when $z = 0$.

where

$$K(k) = \int_0^1 \frac{\mathrm{d}t}{\sqrt{1-k^2t^2}\sqrt{1-t^2}}, \quad E(k) = \int_0^1 \frac{\sqrt{1-k^2t^2}}{\sqrt{1-t^2}}\mathrm{d}t. \tag{9.12}$$

We have checked that the function $A_n(\rho, 0)$ in Eq. (9.9) tends to $A_c(\rho, 0)$ in Eq. (9.11), as n approaches infinity. Also, by taking $\boldsymbol{B} = \nabla \times \boldsymbol{A}$ and suitable variable redefinitions, we recover the corresponding magnetic field [3].

9.3 Calculating the flux

We recall two points mentioned in the introduction. Because the fields diverge at the position of the wires, we will take the flux in a curve similar to the original wire but scaled down by some amount, as in Fig. 9.2(b). We may think of this as a cutoff introduced by the finite width of the wire, or as the situation faced in calculating the flux through a second loop, similar to (but smaller than) the current-carrying one. Also, because the direct calculation of the flux of $\boldsymbol{B}$ involves a two-dimensional integration, we will use Eq. (9.3) and calculate instead the line integral of $\boldsymbol{A}$.

The simplicity gained in utilizing $\boldsymbol{A}$ is particularly striking in the case of the circular current loop, since Eq. (9.10) means that $\boldsymbol{A}$ is independent of θ. Therefore, choosing an integration circle C_ρ, of radius $\rho \in (0, R)$, we find

$$\begin{aligned} \frac{4\pi}{\mu_0 i P}\Phi_c &= \frac{4\pi}{\mu_0 i P}\int_{C_\rho} \boldsymbol{A}\cdot \mathrm{d}\boldsymbol{l} = \frac{4\pi}{\mu_0 i P} A(\rho,0)\, 2\pi\rho \\ &= \frac{4\pi}{\rho+R}\left[(\rho^2+R^2)\, K\left(\frac{2\sqrt{\rho R}}{\rho+R}\right) - (\rho+R)^2\, E\left(\frac{2\sqrt{\rho R}}{\rho+R}\right)\right], \end{aligned} \tag{9.13}$$

where, in going to the second line, we have made ρ and R dimensionless by scaling them by the perimeter P[3]. It is instructive to compare the trivial reasoning on the first line of Eq. (9.13) with what would be needed to calculate the flux directly from the ρ-dependent $\boldsymbol{B}_c$.

Next, we consider the magnetic field produced by a polygon with perimeter P, n equal sides, and carrying the current i. The distance from the center to each of the sides is given by d_n. Consider also a second n-sided polygon C_n whose sides lie a distance $x_{\text{cut}} \in (0, d_n)$ from the same center. The flux through

[3]We have made the variable substitutions $\rho' = \rho/P$ and $R' = R/P = 1/(2\pi)$, and then dropped the primes.

this polygon is given by

$$\begin{aligned}\Phi_n &= \int_{C_n} \boldsymbol{A}_n \cdot \mathrm{d}\boldsymbol{l} = n \int_{\text{first side}} \boldsymbol{A}_n \cdot \mathrm{d}\boldsymbol{l} \\ &= n \int_{-x_{\text{cut}} \tan(\pi/n)}^{x_{\text{cut}} \tan(\pi/n)} (A_n)_y (x_{\text{cut}}, y)\, \mathrm{d}y. \end{aligned} \tag{9.14}$$

Looking back at Eq. (9.6), one notices the need for integrals involving the logarithm of rather complicated functions. Things can be greatly simplified, however. We start by rescaling all distances by the perimeter P, thus rendering the variables x, y, s_n, and d_n appearing in Eq. (9.6) dimensionless[4]. Next we introduce new parameters u and new variables v through

$$\begin{aligned} u &= x_{\text{cut}} - X_{nk}(\pm s_n/2), \\ v &= y - Y_{nk}(\pm s_n/2), \end{aligned} \tag{9.15}$$

for use in Eqs. (9.6) and (9.8). Thus, for Eq. (9.6) we need

$$I_{nk}(u,v) \equiv \int \ln\left[u \sin\beta_{nk} - v\cos\beta_{nk} + \sqrt{u^2+v^2}\right] \mathrm{d}v\,. \tag{9.16}$$

We find[5]

$$I_{nk}[u,v] = \begin{cases} v\,\ln\left(-v+\sqrt{u^2+v^2}\right) + \sqrt{u^2+v^2} & \text{if } \beta_{nk}=0 \\ v\,\ln\left(v+\sqrt{u^2+v^2}\right) - \sqrt{u^2+v^2} & \text{if } \beta_{nk}=\pi \\ -v + u\,\csc\beta_{nk}\,\ln\left(v+\sqrt{u^2+v^2}\right) & \\ \quad +\,(v+u\cot\beta_{nk})\ln\big(u\sin\beta_{nk} & \\ \quad\quad -v\cos\beta_{nk}+\sqrt{u^2+v^2}\big) & \text{otherwise}\,. \end{cases} \tag{9.17}$$

Combining this with Eqs. (9.6)–(9.9), and substituting into Eq. (9.14), one obtains

$$\frac{4\pi}{\mu_0 i P}\Phi_n = n\sum_{k=1}^{n} \cos\beta_{nk}\left(I_{nk}^{+} - I_{nk}^{-}\right), \tag{9.18}$$

where

$$\begin{aligned} I_{nk}^{\pm} = {}& I\left[x_{\text{cut}} - X_{nk}(\pm s_n/2),\ x_{\text{cut}}\tan(\pi/n) - Y_{nk}(\pm s_n/2)\right] \\ & - I\left[x_{\text{cut}} - X_{nk}(\pm s_n/2),\ -x_{\text{cut}}\tan(\pi/n) - Y_{nk}(\pm s_n/2)\right]. \end{aligned} \tag{9.19}$$

We have checked Eqs. (9.13) and (9.18) in two important limits. First, expanding around $x_{\text{cut}} = 0$, we find that the fluxes tend to the product of

---

[4] We have made the variable substitutions $x' = x/P$, $y' = y/P$, $s_n' = s_n/P = 1/n$, and $d_n' = d_n/P = \cot(\theta_n/2)/(2n)$, and then dropped the primes.

[5] We are very grateful to Ana C. Barroso for help with this integral.

the magnetic field at the center with the area of a small central region whose distance to the sides is x_{cut}. Indeed,

$$\Phi_c \;\rightarrow\; B_c^{\text{center}}\, \pi\, x_{\text{cut}}^2 + O(x_{\text{cut}}^3) \;\rightarrow\; 4\pi^3\, x_{\text{cut}}^2\,, \tag{9.20}$$

$$\begin{aligned}\Phi_n \;&\rightarrow\; B_n^{\text{center}}\, n \tan(\pi/n)\, x_{\text{cut}}^2 + O(x_{\text{cut}}^3)\\ &\rightarrow\; 4n^3 \tan^2(\pi/n) \sin(\pi/n)\, x_{\text{cut}}^2\,.\end{aligned} \tag{9.21}$$

Here and henceforth (including in all figures), we normalize the fluxes by $\mu_0 i P/(4\pi)$, the magnetic fields by $\mu_0 i/(4\pi P)$, and we continue to scale all distances by P. Naturally, we can recover Eq. (9.20) from Eq. (9.21) in the limit of n going to infinity. Second, Φ_n tends to Φ_c as n goes to infinity, for all values of x_{cut}. This can be seen in Fig. 9.3, which displays Φ_n for $n = 3$, 4, 8, and Φ_c as a function of x_{cut}. Each flux Φ_n diverges at $x_{\text{cut}} = d_n$, while

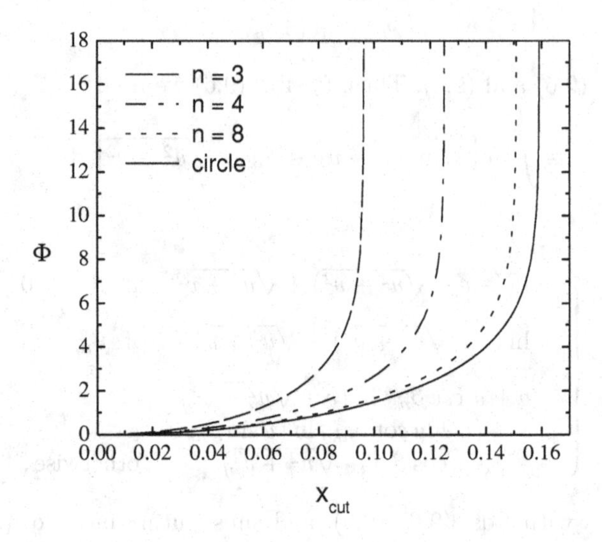

FIGURE 9.3
Auto-fluxes as a function of x_{cut}, for current-carrying polygons with $n = 3$, 4, 8, and for the circular loop.

Φ_c diverges at $x_{\text{cut}} = R$, providing a nontrivial crosscheck on our expressions. Notice that, for each value of $x_{\text{cut}} < d_3$, the curve for Φ_c lies below all other fluxes. Although the fields $\boldsymbol{B}$ vary as one moves away from the center, a very rough way of understanding this result is the following: the field at the center B_n^{center} decreases as n increases—*c.f.* Eq. (9.1); on the other hand, for fixed x_{cut}, the areas through which the flux is being considered are given by $n \tan(\pi/n)\, x_{\text{cut}}^2$, for Φ_n, and by $\pi\, x_{\text{cut}}^2$, for Φ_c, which also decrease as n increases. Therefore, in this case the "area factors" do not compensate for the smaller fields, as seen in Eqs. (9.20) and (9.21).

Since the fluxes diverge for $x_{\text{cut}} = d_n$, we may choose to consider another situation. We take all wires to be of a fixed width δ (in units of P), and we regularize the fluxes by integrating only up to $\rho = R - \delta$, for the circle, and $x_{\text{cut}} = d_n - \delta$, for the polygons. The results are displayed in Fig. 9.4 as a function of δ, for n = 3, 4, 8, and for the circle. We notice the following

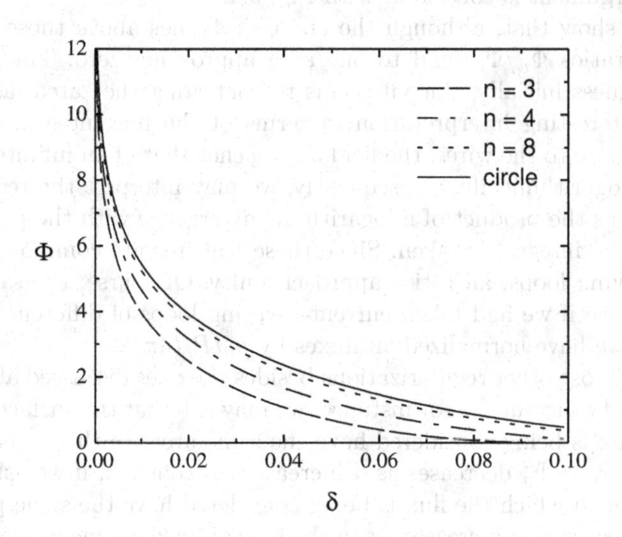

FIGURE 9.4
Auto-fluxes as a function of the width of the wire, for current-carrying polygons with n = 3, 4, 8, and for the circular loop.

features: i) for any finite value of δ, the auto-flux increases as n increases—this indicates that, here, the "area factor" is making up for the smaller value of the magnetic field at the center; ii) again, the curves of Φ_n tend to Φ_c as n increases; iii) the flux diverges as the width of the wires tends to zero, as expected.

Comparing Fig. 9.3 and Fig. 9.4, we notice that Φ decreases with n in the first case, while it increases with n in the second. So, in contrast to the previous case, here the "area factor" compensates for the smaller fields. We can get a rough understanding for this in the following way: for fixed δ, the areas through which the flux is being considered are given by

$$n \tan(\pi/n)\,(d_n - \delta)^2 = \frac{1}{4n}\cot(\pi/n) - \delta + n\tan(\pi/n)\,\delta^2, \tag{9.22}$$

for Φ_n, and by

$$\pi(R - \delta)^2 = \frac{1}{4\pi} - \delta + \pi\,\delta^2, \tag{9.23}$$

for Φ_c, in units of P^2. As δ vanishes, the areas in Eqs. (9.22) and (9.23) are dominated by their first terms, which do increase enough as to offset the order of the field magnitudes. Of course, this is a very crude argument, since, because the fields vary in different ways as one moves away from the center, using B_n^{center} in the reasoning is a considerable source of error. Nevertheless, this rough argument is consistent with Fig. 9.4.

One can show that, although the curve of Φ_c lies above those of Φ_n for $\delta \neq 0$, the ratios Φ_n/Φ_c tend to one as δ approaches zero. This might be difficult to guess initially, since it seems to contradict the "area factor", but it has an interesting interpretation in terms of the line integral of $\boldsymbol{A}$. For points very close to the wires, the field approaches that of an infinite wire and $\boldsymbol{A}$ diverges logarithmically. Consequently, we may interpret the result of the line integral as the product of a logarithmic divergence with the perimeter P over which the integral is taken. Since these features are common to all the current-carrying loops, all ratios approach unity. Of course, the same would not be the case if we had taken current-carrying loops of different perimeter (recall that we have normalized all fluxes by $\mu_0 i P/(4\pi)$).

We can choose other regularizations besides the ones discussed above (constant x_{cut} and constant δ). For instance, we may ask that the surfaces through which the flux is being considered have the same area. In this case, as in the case of fixed x_{cut}, Φ_n decreases as n increases. In contrast, if we ask that the surfaces through which the flux is being considered have the same perimeter, then Φ_n increases as n increases, as in the case of fixed δ. One can get a rough understanding for these features along the lines of the analysis made above.

Finally, we recall that the line integrals of $\boldsymbol{A}$ have been performed over curves C_n and C_ρ identical to the current-carrying wires, but smaller. This is what one needs for the calculation of the mutual inductance between two (polygonal or circular) current-carrying wires of equal shape and different scales that lie on the same plane and are concentric. Our results apply directly to that case.

9.4 Conclusions

Motivated by a simple exercise in elementary electromagnetism, we have studied the interplay between the magnetic fields and the areas of current-carrying polygonal and circular wires of equal perimeter. We have calculated the vector potential $\boldsymbol{A}$ for these situations, because its line integral provides a much simpler way of computing the magnetic fluxes; this example illustrates the usefulness of $\boldsymbol{A}$ in practical calculations. Since the corresponding auto-fluxes diverge, we have discussed a number of regularizations, comparing the fluxes in each case, and seeking intuitive arguments for the results. As a bonus, our results can be applied directly to the calculation of mutual inductances in a variety of situations.

Acknowledgments

We are very grateful to Ana C. Barroso for considerable help with some integrations, to A. Nunes for reading and commenting on this manuscript, and to our students for their prodding questions.

References

[1] See, for example, D. Halliday, R. Resnick, and J. Walker, *Fundamentals of Physics*, extended 6th ed. (John Wiley and Sons, New York, 2001) p. 706.

[2] M. I. Grivich and D. P. Jackson, "The magnetic field of current-carrying polygons: An application of vector field rotations," Am. J. Phys. **68**, 469–474 (1999).

[3] H. Erlichson, "The magnetic field of a circular turn," Am. J. Phys. **57**, 607–610 (1989).

[4] J. A. Miranda, "Magnetic field calculation for arbitrarily shaped planar wires," Am. J. Phys. **68**, 254–258 (1999). For a very interesting extension of the techniques used for current-carrying planar wires into electrostatic problems, see M. H. Oliveira and J. A. Miranda, "Biot-Savart-like law in electrostatics," Eur. J. Phys. **22**, 31–38 (2001).

10

The Elastic Bounces of a Sphere between Two Parallel Walls

J. M. Tavares

CONTENTS

The kinematics of a sphere of radius R and mass m bouncing elastically in a horizontal channel is studied in detail. Explicit expressions for the position and linear and angular velocities of the sphere at any collision are obtained. It is shown that if the moment of inertia $I = \gamma m R^2$ is such that $(1-\gamma)/(1+\gamma) = \cos 2\pi/k$ with k an integer and $k \geq 4$, the motion is periodic, with repetition after k collisions for k even, and after $2k$ collisions for k odd. The motion of an homogeneous sphere (which is not periodic) is analyzed by investigating the mean properties of its velocities and positions. The analogy between this motion and that of a periodic oscillator is discussed.

10.1 Introduction

The study of the elastic collisions of a sphere with a rough surface has been a source of non-intuitive results. Phenomena like backward reflection, vertical reflection [1], repetition of motions after certain reflections [2] and endless

Reproduced from J. M. Tavares, "The elastic bounces of a sphere between two parallel walls," American Journal of Physics **75**, 690–695 (2007), https://doi.org/10.1119/1.2742402, with the permission of the American Association of Physics Teachers.

bounces between two vertical (and finite) surfaces [3] are a few examples of unexpected possible motions. All of these results are direct consequences of a few simple hypotheses and conservation of angular momentum and conservation of energy. Moreover, most of the results can be easily verified experimentally (at least qualitatively) using a superball.

In this paper, I study the kinematics of a sphere bouncing elastically between two horizontal surfaces in the absence of gravity (the horizontal channel in Ref. [3]). I will focus on the role of the moment of inertia to show that for specific mass distributions, there are periodic motions in which the positions and velocities repeat after an even number of collisions (≥ 4). To characterize the general case for which the motion is not periodic, I perform a simple statistical analysis for a homogeneous sphere.

The paper is organized as follows. In Sec. 10.2, the elastic reflection of a sphere on a rough surface is briefly reviewed following Refs. [1,2]. In Sec. 10.3, the case of consecutive elastic reflections in a horizontal channel is developed as in Ref. [3], but extended to obtain explicit results for the velocities and positions of the moving sphere at any collision. Finally, the results obtained for the kinematics of spheres for several values of the moment of inertia are discussed.

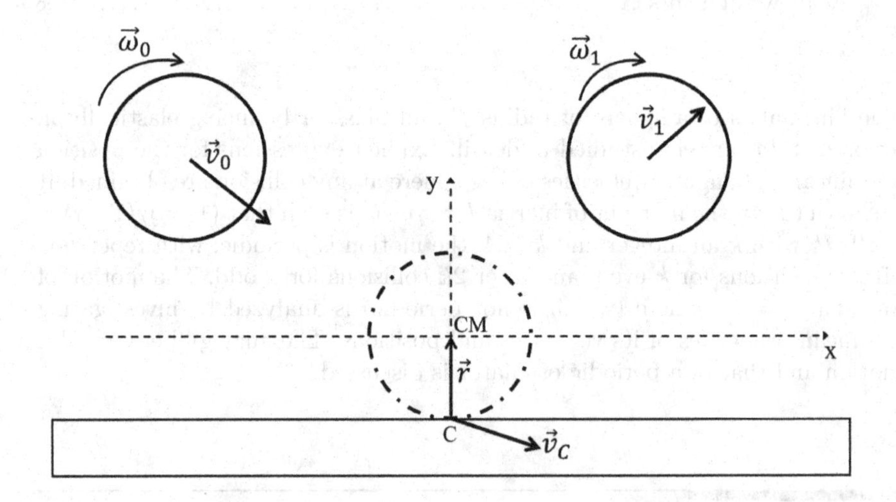

FIGURE 10.1
The collision of a sphere with a horizontal surface. The plane of the figure is perpendicular to the surface and is defined by $\vec{r}$, the vector joining the contact point C with the center of mass of the sphere, and $\vec{v}_C$, the velocity of the contact point. The contact force (not shown) is in this plane. The y axis defines the normal or perpendicular direction, and the x axis defines the tangential or horizontal direction.

10.2 Collision with a horizontal wall

Consider the situation depicted in Fig. 10.1. A sphere of mass m, radius R, moment of inertia $I = \gamma m R^2$, whose geometrical center coincides with the center of mass, has, immediately before colliding with a plane wall, a center of mass velocity $\vec{v}_0$ and angular velocity (or spin) $\vec{\omega}_0$. Immediately after the collision, the velocity of the sphere is $\vec{v}_1$ and the spin is $\vec{\omega}_1$.

We assume (1) conservation of kinetic energy (elastic collision), (2) the contact force acts only at the contact point C and is contained in the plane defined by the center of mass and the velocity of C (see Fig. 10.1), and (3) the inversion of the normal component of the velocity by the collision. Then as shown in Ref. [1], it is possible to obtain the final velocities $\vec{v}_1$ and $\vec{\omega}_1$ as a function of the initial velocities $\vec{v}_0$ and $\vec{\omega}_0$. In the reference frame of Fig. 10.1, assumption (3) is equivalent to

$$v_{1,y} = -v_{0,y}. \tag{10.1}$$

The assumption made for the contact force has two consequences: the angular momentum of the sphere about the contact point is conserved,

$$-v_{0,x} + \gamma R \omega_{0,z} = -v_{1,x} + \gamma R \omega_{1,z}, \tag{10.2}$$

and the z component of the velocity and the x and y components of the angular velocity remain unchanged. For simplicity, we will assume that $v_{0,z} = \omega_{0,x} = \omega_{0,y} = 0$ [1]. As a consequence of (2) and (3), conservation of kinetic energy is expressed by

$$v_{0,x}^2 + \gamma R^2 \omega_{0,z}^2 = v_{1,x}^2 + \gamma R^2 \omega_{1,z}^2. \tag{10.3}$$

The final velocities of the sphere $(\vec{v}_1, \vec{\omega}_1)$ can be calculated as a function of the initial velocities $(\vec{v}_0, \vec{\omega}_0)$ using Eqs. (10.1)–(10.3). This system of equations has two solutions. The first one is Eq. (10.1) plus $v_{1,x} = v_{0,x}$ and $\omega_{1,z} = \omega_{0,z}$, and corresponds either to the elastic collision of a point particle with a wall or to assuming that the contact force between the sphere and the wall is perpendicular to the latter. In this case, the rotation of the sphere is irrelevant.

The second solution corresponds to an elastic collision with a rough surface so that the contact force also has a component parallel to the wall. As originally proposed in Ref. [2], this solution can be represented with the aid of a collision matrix $\tilde{M}$. The normal component of the final velocity is obtained from Eq. (10.1). The tangential component of the velocity and the angular velocity can be obtained through

$$\tilde{v}_1 = \tilde{M} \tilde{v}_0, \tag{10.4}$$

[1] Within this simplification, the developments that follow may be applied to other objects with cylindrical symmetry (for example, rings and cylinders) whose center of mass coincides with their geometrical center, whose cross-section in the Oxy plane is a circle or a circumference, and that rotate around the z-axis.

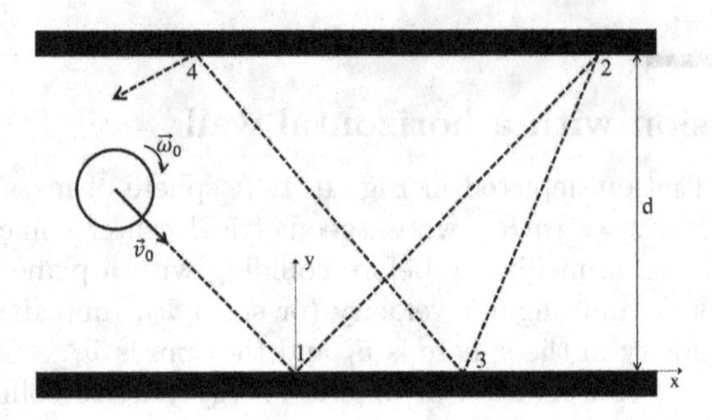

FIGURE 10.2
A sphere bounces between two parallel horizontal surfaces or walls (horizontal channel). There is no gravity. The sphere begins its motion with velocity $\vec{v}_0$ and $\vec{\omega}_0$ and then performs successive collisions with both walls. These collisions are successively numbered as shown. The velocities just after collision n are denoted by $\vec{v}_n$ and $\vec{\omega}_n$.

where $\tilde{v}_i$ represents the velocities in matrix form,

$$\tilde{v}_i = \begin{bmatrix} v_{i,x} \\ \bar{\omega}_{i,z} \end{bmatrix}, \tag{10.5}$$

$\tilde{M}$ is the collision matrix,

$$\tilde{M} = \frac{1}{1+\gamma} \begin{bmatrix} 1-\gamma & -2\gamma \\ -2 & -(1-\gamma) \end{bmatrix}, \tag{10.6}$$

and $\bar{\omega}_{i,z} \equiv R\omega_{i,z}$ is the linear velocity of a point of the surface of the sphere relative to its center of mass.

10.3 Successive elastic collisions of a sphere with two parallel planar walls

Let us now consider the situation depicted in Fig. 10.2, where a sphere bounces repeatedly between two parallel and planar walls placed at a distance d from each other. Before the first collision the sphere has velocities $(\vec{v}_0, \vec{\omega}_0)$. After n collisions the velocities of the sphere will be $(\vec{v}_n, \vec{\omega}_n)$. We will show that it is possible to obtain an explicit expression that allows the calculation of $(\vec{v}_n, \vec{\omega}_n)$ from $(\vec{v}_0, \vec{\omega}_0)$.

The y component of the velocity of the center of mass after collision n is easily calculated by considering the successive application of Eq. (10.1),

$$v_{n,y} = -v_{n-1,y} = (-1)^n v_{0,y}, \tag{10.7}$$

which means that $v_{n,y}$ will repeat every two collisions.

In the reference frame of Fig. 10.2, Eq. (10.4) can be generalized to any collision with the lower wall (n odd),

$$\tilde{v}_n = \tilde{M}\tilde{v}_{n-1}. \tag{10.8}$$

The matrix $\tilde{M}$ cannot be directly used to obtain the final velocities after a collision with the upper wall (n even). The collision matrix $\tilde{M}$ is determined by the choice of a reference frame where the y axis points out of the wall (see Figs. 10.1 and 10.2). If we use the reference frame of Fig. 10.2 to study the collision with the upper wall, the only change in Eqs. (10.1)–(10.3) is in the equation for the conservation of angular momentum. The sign of the z component of the angular velocity must be reversed to obtain

$$-v_{n-1,x} - \gamma R\omega_{n-1,z} = -v_{n,x} - \gamma R\omega_{n,z}, \tag{10.9}$$

where n is even. As a consequence, when n is even, the collision matrix $\tilde{M}$ can be applied to the velocities $\tilde{v}_{n-1}$ with the component $\bar{\omega}_{n-1,z}$ reversed to obtain $\tilde{v}_n$ with the component $\bar{\omega}_{n,z}$ reversed. In matrix notation, we have for n even,

$$\tilde{D}\tilde{v}_n = \tilde{M}\tilde{D}\tilde{v}_{n-1}, \tag{10.10}$$

where $\tilde{D}$ is the rotation matrix,

$$\tilde{D} = \begin{bmatrix} 1 & 0 \\ 0 & -1 \end{bmatrix}. \tag{10.11}$$

Therefore, for n even, $\tilde{v}_n$ can be obtained from $\tilde{v}_{n-1}$ through

$$\tilde{v}_n = \tilde{D}^{-1}\tilde{M}\tilde{D}\tilde{v}_{n-1}. \tag{10.12}$$

To obtain an explicit expression for $\tilde{v}_n$ as a function of $\tilde{v}_0$ we have to combine Eqs. (10.8) and (10.12). Because $\tilde{D}$ is a unitary matrix ($\tilde{D}^2 = \tilde{I}$, where $\tilde{I}$ is the unit matrix, and $\tilde{D}^{-1} = \tilde{D}$), Eqs. (10.8) and (10.12) can be written in the form,

$$\tilde{v}_n = \begin{cases} \tilde{D}\tilde{M}\tilde{D}\tilde{M}\ldots\tilde{D}\tilde{M}\tilde{v}_0 & \text{if } n \text{ is even} \\ \tilde{D}\tilde{D}\tilde{M}\tilde{D}\tilde{M}\ldots\tilde{D}\tilde{M}\tilde{v}_0 & \text{if } n \text{ is odd} \end{cases}. \tag{10.13}$$

If we use,

$$\tilde{D}^n = \begin{bmatrix} 1 & 0 \\ 0 & (-1)^n \end{bmatrix}, \tag{10.14}$$

Eq. (10.13) can be put in a form applicable for even and odd n,

$$\tilde{v}_n = \tilde{D}^n \left(\tilde{D}\tilde{M}\right)^n \tilde{v}_0. \tag{10.15}$$

For calculational purposes, Eq. (10.15) (already obtained in Ref. [3]) is simple enough to calculate the successive velocities of the bouncing sphere. In Ref. [3] a trial solution for $\tilde{v}_n$ was proposed. Here we obtain the analytical solution explicitly by diagonalizing the matrix $\tilde{D}\tilde{M}$,

$$\tilde{D}\tilde{M} = \frac{1}{1+\gamma}\begin{bmatrix} 1-\gamma & -2\gamma \\ 2 & 1-\gamma \end{bmatrix}. \tag{10.16}$$

The (complex) eigenvalues of this matrix are $(1-\gamma \pm 2\sqrt{\gamma}i)/(1+\gamma)$; the eigenvectors can be represented as the columns of the matrix $\tilde{P}$,

$$\tilde{P} = \begin{bmatrix} 1 & i\sqrt{\gamma} \\ i/\sqrt{\gamma} & 1 \end{bmatrix}, \tag{10.17}$$

Therefore, the matrix $\tilde{D}\tilde{M}$ can be rewritten as

$$\tilde{D}\tilde{M} = \tilde{P}\tilde{\Lambda}\tilde{P}^{-1}, \tag{10.18}$$

where

$$\tilde{\Lambda} = \frac{1}{1+\gamma}\begin{bmatrix} 1-\gamma-2\sqrt{\gamma}i & 0 \\ 0 & 1-\gamma+2\sqrt{\gamma}i \end{bmatrix}. \tag{10.19}$$

The complex eigenvalues of $\tilde{D}\tilde{M}$ can be expressed in the geometrical form,

$$\frac{1-\gamma \pm 2\sqrt{\gamma}i}{1+\gamma} = e^{\pm i\theta} \tag{10.20}$$

with

$$\cos\theta = \frac{1-\gamma}{1+\gamma} \text{ and } \sin\theta = \frac{2\sqrt{\gamma}}{1+\gamma}. \tag{10.21}$$

Finally, Eq. (10.15) can be expressed in the simpler form

$$\tilde{v}_n = \tilde{D}^n \tilde{P}\tilde{\Lambda}^n \tilde{P}^{-1}\tilde{v}_0, \tag{10.22}$$

where

$$\tilde{\Lambda}^n = \begin{bmatrix} e^{-in\theta} & 0 \\ 0 & e^{in\theta} \end{bmatrix}. \tag{10.23}$$

The expansion of Eq. (10.22) using Eqs. (10.17) and (10.23) leads to an explicit expression for $\tilde{v}_n$ as a function of $\tilde{v}_0$,

$$v_{n,x} = \cos(n\theta)v_{0,x} - \sqrt{\gamma}\sin(n\theta)\bar{\omega}_{0,z} \tag{10.24}$$

$$\bar{\omega}_{n,z} = (-1)^n\left(\sin(n\theta)v_{0,x}/\sqrt{\gamma} + \cos(n\theta)\bar{\omega}_{0,z}\right). \tag{10.25}$$

Equations (10.24) and (10.25) can be simplified further by using the equality $a\sin x + b\cos x = \sqrt{a^2+b^2}\sin(x+\phi)$ with $\phi = \arctan(a/b)$ [4],

$$v_{n,x} = A\cos(n\theta + \phi_z) \tag{10.26}$$

$$\sqrt{\gamma}\bar{\omega}_{n,z} = (-1)^n A\sin(n\theta + \phi_z), \tag{10.27}$$

where $A \equiv \sqrt{v_{0,x}^2 + \gamma\bar{\omega}_{0,z}^2}$ and $\phi_z = \arctan\sqrt{\gamma}\bar{\omega}_{0,z}/v_{0,x}$. Equations (10.26) and (10.27) (together with Eq. (10.7)) provide a complete characterization of the motion of a sphere bouncing between two parallel planar walls.

Notice that for particular distributions of the mass of the sphere (that is, for specific values of the parameter γ, or equivalently of θ through Eq. (10.20)), the velocities will repeat after some collisions. If the mass distribution is such that $\theta = 2\pi/k$, with k an integer and ≥ 4,[2] then due to the dependence on $n\theta$ through the sinusoidal functions and the factor $(-1)^n$ in Eq. (10.27), we have that the component $v_{n,x}$ will repeat after k collisions; the component $\bar{\omega}_{n,z}$ will repeat after $2k$ collisions for k odd, after $k/2$ collisions for k even and $k/2$ odd, and after k collisions for all other cases.

To illustrate these periodicities, the velocities given by Eqs. (10.26) and (10.27) are plotted in Figs. 10.3(a) and (b) for mass distributions corresponding to $k = 4$, 5, 6. The initial condition was $\bar{\omega}_{0,z} = 0$. For $k = 4$ (or $\gamma = 1$, a ring), both $v_{n,x}$ and $\bar{\omega}_{n,z}$ repeat after 4 collisions. Due to the initial condition, either $v_{n,x}$ or $\bar{\omega}_{n,z}$ are 0 for each n, meaning that in each collision, the kinetic energy is completely transferred between the two corresponding degrees of freedom. The case $k = 5$ illustrates the general behavior obtained for k odd: $v_{n,x}$ repeats after 5 collisions, but $\bar{\omega}_{0,z}$ has a period of 10 collisions. Note that $k = 5$ corresponds to $\gamma \approx 0.53$ and therefore to a mass distribution in between two typical situations: a homogeneous sphere ($\gamma = 2/5$) and a spherical shell ($\gamma = 2/3$). The case $k = 6$ ($\gamma = 1/3$), in which $v_n(x)$ repeats after 6 collisions and $\bar{\omega}_{n,z}$ repeats after 3 collisions, exemplifies the periodic variation of the velocities when $k = 2m$ with m odd. For other choices of the initial conditions only the values of the velocities will change, not their periodicity.

There are several objects with cylindrical symmetry and moments of inertia corresponding to $k \geq 5$. One possibility is to consider a disk of radius βR ($\beta \leq 1$) and mass density ρ, surrounded by a circular layer of radius R and density $\alpha\rho$. By adjusting the parameters α and β all values of k can be obtained. The limit $k \to \infty$ (that is, $\gamma \to 0$) is achieved by taking $\alpha \to 0$ and $\beta \to 0$ (that is, by considering all the mass placed in the center of the object).

In Figs. 10.3(c) and 10.3(d), the velocities $v_{n,x}$ and $\bar{\omega}_{n,z}$ are shown for $\gamma = 2/5$ ($\theta = \arccos(3/7) \approx 1.13$). This example illustrates the most general case where θ is not equal to $2\pi/k$ and therefore the velocities are not periodic. The sinusoidal dependence of the velocities manifests itself in their (aperiodic) oscillations between positive and negative values that, as will be shown, confine the sphere to bounce in a limited region of the horizontal channel.

The trajectory of the center of mass can be explicitly calculated from Eqs. (10.7) and (10.26). Because there are no forces acting on the sphere between the collisions, the trajectory of the center of mass is the line that links the successive positions where the collisions occur. If y_1 is the vertical coordinate of the center of mass at collision 1 and d is the width of the horizontal

[2]Because $\cos\theta = (1-\gamma)/(1+\gamma)$, and $0 < \gamma \leq 1$, we must have $0 < \theta \leq \pi/2$, and therefore $k \geq 4$.

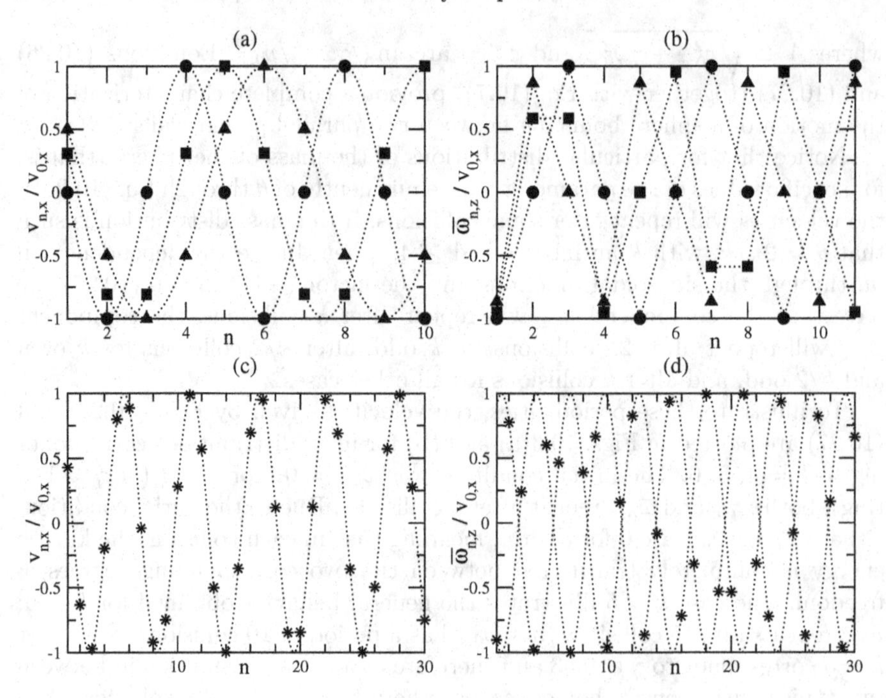

FIGURE 10.3
(a) and (c) Values of $v_{n,x}/v_{0,x}$ and (b) and (d) $\bar{\omega}_{n,z}/v_{0,x}$ as given by Eqs. (10.24) and (10.25). The initial condition is $\bar{\omega}_{0,z} = 0$. In (a) and (b) $\theta = 2\pi/k$ for $k = 4$ (circles), $k = 5$ (squares), and $k = 6$ (triangles). In (c) and (d) $\theta = \arccos(1-\gamma)/(1+\gamma)$ with $\gamma = 2/5$, which corresponds to a homogeneous sphere. (a) and (c) The dashed lines are the continuous representation of the functions in Eq. (10.24) and (d) of the two branches (n odd and even) of Eq. (10.25); in (b) the dashed lines link the successive values of the angular velocity. Only the points have physical meaning.

channel, then the vertical coordinate at collision n is

$$y_n = y_1 + \frac{1 + (-1)^n}{2}(d - 2R). \tag{10.28}$$

The x coordinate of the center of mass at collision n can be determined using the recursion relation, $x_n = x_{n-1} + v_{n-1,x}\Delta t$, where $\Delta t = (d - 2R)/|v_{0,y}|$ is the time between two successive collisions. If the center of mass in collision 1 is at x_1, then at collision n it will be at

$$x_n = x_1 + \sum_{i=1}^{n-1} v_{i,x}\Delta t. \tag{10.29}$$

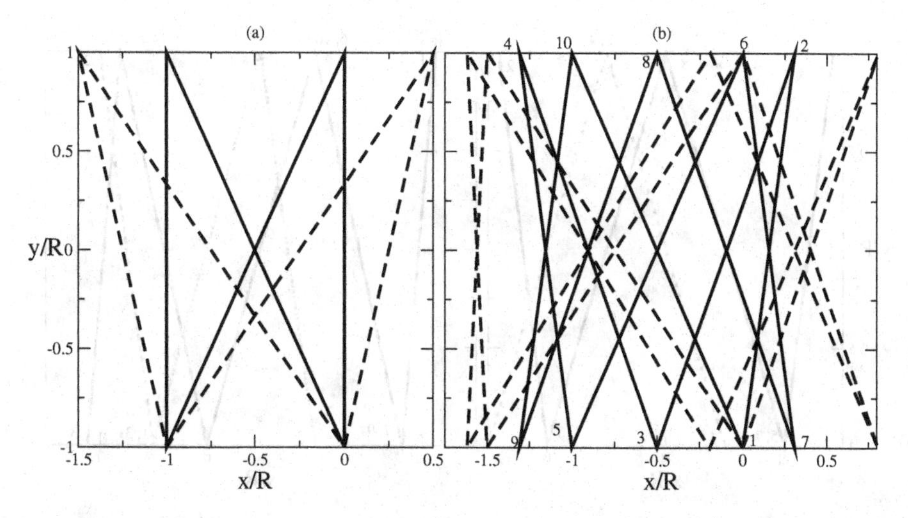

FIGURE 10.4
Representation of periodic trajectories of the center of mass of spheres bouncing between two parallel walls: (a) $k = 4$ (or $\gamma = 1$, a ring); (b) $k = 5$ (or $\gamma \approx 0.53$). The distance between the walls is $d = 4R$. The first collision takes place at $(x_1, y_1) = (0, -R)$. The velocities before the first collision are $v_{0,x} = |v_{0,y}|$ and $\bar{\omega}_{0,z} = 0$ (solid lines); $v_{0,x} = 1.5|v_{0,y}|$ and $\bar{\omega}_{0,z} = -0.5|v_{0,y}|$ (dashed line). The integer numbers in (b) are the indexes of successive collisions for the trajectory represented as a solid line.

By using Eq. (10.26) and the equality [4],

$$\sum_{k=0}^{n} \cos(k\theta + \phi) = \frac{\sin\left(\frac{n+1}{2}\theta\right)\cos\left(\phi + \frac{n}{2}\theta\right)}{\sin\frac{\theta}{2}}, \tag{10.30}$$

it is possible to obtain an explicit expression for x_n,

$$x_n = x_1 + \frac{A\Delta t}{2}\left[\sin(n\theta + \phi_z - \theta/2) - \sin(\phi_z + \theta/2)\right]. \tag{10.31}$$

Equation (10.31) shows that for every value of θ the motion is confined to the region of the channel in which $-1 \leq (x - x_1)\frac{2}{A\Delta t} + \sin(\phi_z + \theta/2) \leq 1$.

An analysis of Eqs. (10.28) and (10.31) reveals that for $\theta = 2\pi/k$ ($k \geq 4$) the center of mass will have periodic trajectories that repeat each k collisions for k even and after $2k$ collisions for k odd. In Fig. 10.4, we plot four examples of these trajectories for $k = 4$ and $k = 5$. The detailed form of the trajectories depends on the initial condition. If k is odd, $x_{n+k} = x_k$ and $y_{n+k} = -y_n$; that is, after k collisions the sphere will hit the opposite wall at precisely the same x coordinate.

An appropriate choice of the initial conditions for k even can lead to trajectories where each line connecting the two walls repeats twice during each

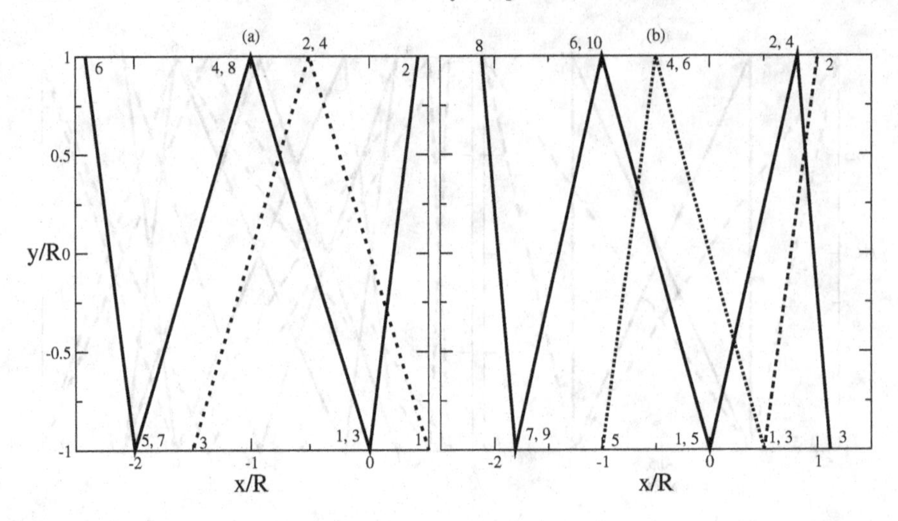

FIGURE 10.5
Periodic trajectories of the center of mass of spheres in a horizontal channel, in the case where each line repeats twice in a period. (a) The initial condition is $v_{0,x} = \bar{\omega}_{0,z}$ (or $\phi_z = \arctan\sqrt{\gamma}$); (b) the initial condition is $\bar{\omega}_{0,z} = 0$ (or $\phi_z = 0$). The dashed line corresponds to $k = 4$ in (a) and to $k = 6$ in (b); the solid line corresponds to $k = 8$ in (a) and to $k = 10$ in (b). The integer numbers are the indexes of successive collisions. Before the first collision the sphere is moving to the right ($v_{0,x} > 0$).

period. In Fig. 10.5, we display examples of these trajectories. These initial conditions can be obtained using Eq. (10.31) to search for the values of m and n that satisfy $x_n - x_{n+m} = 0$, with $\theta = 2\pi/k$, m even (to ensure that $y_n = y_{n+m}$) and $m < k$ (that is, repetitions of the position within a period). By solving this relation, we conclude that these repetitions occur if (a) $k = 4, 8, \ldots, v_{0,x} = \bar{\omega}_{0,z}$ (or $\phi_z = \arctan\sqrt{\gamma}$) and $2(2n+m)/k$ is an integer (see Fig. 10.5(a)); or if (b) $k = 6, 10, \ldots, \bar{\omega}_{0,z} = 0$ (or $\phi_z = 0$) and $2(2n+m-1)/k$ is an integer (see Fig. 10.5(b)). For example, when $k = 10$, the collisions of index 1 and 5, 2 and 4, 7 and 9, and 6 and 10 will take place at the same positions, if the initial condition $\bar{\omega}_{0,z} = 0$ is chosen.

The trajectory of a homogeneous sphere is, as can be concluded from Eq. (10.31), bounded but not periodic. In Ref. [3] this motion was studied for the first few collisions. We can take advantage of the explicit expressions (10.26)–(10.29) obtained for the velocities and positions and study the mean properties of the motion after a large number of collisions. In Fig. 10.6(a) a trajectory of the center of mass of a homogeneous sphere after 300 collisions is plotted. We can clearly see that there is a tendency to follow some paths and to hit the wall at points close to each other. Figure 10.6(b) shows that there are some regions of the walls that still have not been hit after 300 collisions. However, after 50000 collisions all the regions have been hit, and there is a much larger probability to hit the regions near the boundaries of the motion than any other inner region of the same size.

In Figs. 10.6(c) and 10.6(d) an analysis of the mean velocities at the exit of a collision that takes place at x/R is presented. These mean velocities are calculated in the following way. For a given number of collisions, the coordinates of the position at which each collision occurs and the values of v_x and $\bar{\omega}_z$ (and also of v_x^2 and $\bar{\omega}_z^2$) just after the collision are registered. The interval of x over which collisions have occurred is divided into 50 equal subintervals (for each wall of the channel). Simple averages of v_x, $\bar{\omega}_z$, v_x^2 and $\bar{\omega}_z^2$ are then performed for the collisions that occurred in each subinterval. In this way we obtain a representation of the mean velocities and mean square velocities ($<v_x>(x)$, $<v_x^2>(x)$, $<\bar{\omega}_z>(x)$, $<\bar{\omega}_z^2>(x)$) in histograms over the range in x where collisions have occurred.

The mean value of the horizontal velocity of the center of mass of the sphere ($<v_x>$) just after a collision is zero near the center of the interval where the sphere moves, negative in the right half of the interval, and is positive in the left half. In contrast, $\sqrt{<v_x^2>}$, the square root of the mean square of the horizontal velocity, is a maximum at the center of the interval and a minimum at the boundaries of the trajectory. The mean value of the angular velocity $<\bar{\omega}_z>$ is also zero at the center. In collisions with the upper (lower) wall the angular velocity is positive (negative) to the right and negative (positive) to the left of the center of the interval. Also the square root of the mean square of the angular velocity $\sqrt{<\bar{\omega}_z^2>}$ is, for collisions with both walls, a minimum at the center of the trajectory and maximal at the boundaries.

Therefore, at the center of the trajectory it is equally probable to find the sphere just after a collision moving right or left, and rotating to the right or to the left. At these points the rotation velocity is small and the translational velocity is large. At the boundaries it is more probable to find the sphere just after a collision moving toward the center of the trajectory with small translational and angular velocities. The direction of the angular velocity at the exit of a collision at the boundaries depends on whether the sphere has collided with the upper or the lower wall. In any case, this direction is such that the contribution of rotation to the velocity of the point of the sphere that will hit the wall in the next collision is opposite to the contribution of the translational velocity.

The results in Fig. 10.6 show some general features that are independent of the initial conditions and of the value of γ. The initial conditions affect the value of A in Eq. (10.31), that is, the size of the region of the channel that is visited by the sphere. The initial velocities and γ determine the relation between the maximum values of $<v_x^2>$ and $<\bar{\omega}_z^2>$, but not the general shape of the plots.

Some of these general features suggest an analogy with a harmonic oscillator. The quantity $\sqrt{<v_x^2>}$ exhibits a maximum value at the center of the interval of motion in x. The calculated probability of having a collision at x (full line in Fig. 10.6(b)) fits the functional form $1/\sqrt{(x - x_-)(x_+ - x)}$ (not shown in the figure), where x_+ and x_- are, respectively, the largest and the smallest values of the x coordinate of the collisions considered. The probability density of finding a one-dimensional harmonic oscillator at position x is also proportional to this function [5]. However, the mean number of collisions that occur between two successive collisions in a given region of the walls depends strongly on the position of the region considered. For the example in Fig. 10.6, the time between two collisions in the central region is about eight times larger than the time between two collisions in the boundaries. Therefore, this motion has apparently only one characteristic time scale (the time between two successive collisions), except for the cases where $\theta = 2\pi/k$.

We can also study the way the (constant) kinetic energy is distributed over the degrees of freedom of horizontal translation and rotation. For a time much larger than the time between two collisions, the ratio between the mean energies associated with these two degrees of freedom, that is,

$$\lim_{n\to\infty} \frac{\sum_{i=1}^{n} v_{i,x}^2}{\sum_{i=1}^{n} \gamma \bar{\omega}_{i,z}^2}, \tag{10.32}$$

can be calculated from Eqs. (10.26) and (10.27). If we use Eq. (10.30) and the trigonometric identity $\cos^2(x) = (1 + \cos(2x))/2$, we find that this limit is equal to unity. Therefore we find a simple example of the equipartition theorem: the kinetic energy is distributed equally over the degrees of freedom available to it.

The results of this paper can be the basis for student projects. The exploration of the statistical properties of the motion is easy to perform. Students can be challenged to go beyond what is presented in Fig. 10.6 to calculate the mean number of collisions between two collisions at x and to study the dependence of the distributions on the initial conditions. Another interesting project would be animate the motion, with the value of γ as an input parameter.

Acknowledgments

I thank P. I. C. Teixeira for a critical reading of this manuscript.

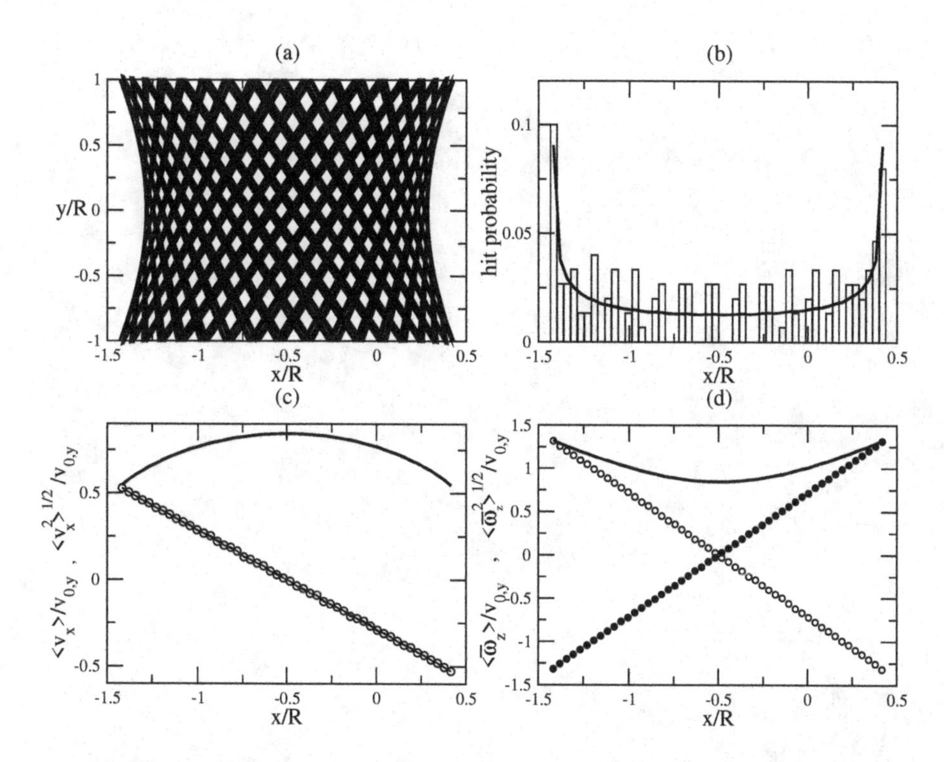

FIGURE 10.6
Characterization of the motion of a homogeneous sphere ($\gamma = 2/5$) between two parallel walls. The initial conditions are $(x_1, y_1) = (0, -R)$, $v_{0,x} = |v_{0,y}|$, and $\bar{\omega}_{0,z} = 0$. (a) Trajectory after 300 collisions. (b) Histogram as a function of x/R of the probability of finding the center of mass of the sphere in a collision with the lower wall, during 300 collisions (bars) and during 50000 collisions (line). The interval in x/R in which the sphere moves was divided into 50 equal subintervals. (c) The histogram after 50000 collisions of the mean values of the horizontal velocity, $<v_x>$ (symbols), and the square root of the mean square horizontal velocity, $\sqrt{<v_x^2>}$ (line), just after a collision with the lower wall that took place when the center of mass was at x/R. (d) Histogram after 50000 collisions of the mean values of the angular velocity, $<\bar{\omega}_z>$ (full (open) symbols—collision with upper (lower) wall), and of the square root of the mean square angular velocity $\sqrt{<\bar{\omega}_z^2>}$ (solid line, collision with lower wall) at the exit of a collision that took place when the center of mass was at x/R.

References

[1] R. L. Garwin, "Kinematics of the ultraelastic rough ball," Am. J. Phys. **37**, 88–92 (1969).

[2] G. L. Strobel, "Matrices and superballs," Am. J. Phys. **36**, 834–837 (1968).

[3] B. T. Hefner, "The kinematics of a superball bouncing between two vertical surfaces," Am. J. Phys. **72**, 875–883 (2004).

[4] I. S. Gradshteyn and I. M. Ryszhik, *Table of Integrals, Series, and Products*, 5th ed. (Academic Press, London 1994), p. 36.

[5] H. J. Pain, *The Physics of Vibrations and Waves*, 2nd ed. (John Wiley and Sons, New York, 1979), p. 42.

11

How Short and Light Can a Simple Pendulum Be for Classroom Use?

V. Oliveira

CONTENTS

We compare the period of oscillation of an ideal simple pendulum with the period of a more "real" pendulum constituted by a rigid sphere and a rigid slender rod. We determine the relative error in calculating the local acceleration of gravity if the period of the ideal pendulum is used instead of the period of this real pendulum.

11.1 Introduction

The simple pendulum is a trivial experiment commonly done by high-school and university students to determine the local acceleration of gravity. A simple pendulum is ideally described as a point mass suspended by a massless rigid rod from some pivot point, about which it is allowed to oscillate. Due to its beauty and simplicity, the simple pendulum is frequently addressed in introductory physics textbooks [1] and has been intensively studied in the literature [2–7]. Nevertheless, most reported studies deal with the extension of the small-angle approximation to large-angle oscillations [4–7], and very little

Used with permission of IOP Publishing Ltd, from V. Oliveira, "How short and light can a simple pendulum be for classroom use?," Physics Education **49**, 387–389 (2014), https://doi.org/10.1088/0031-9120/49/4/387, permission conveyed through Copyright Clearance Center, Inc.

DOI: 10.1201/9781003187103-11

attention has been given to the physical characteristics of a real classroom pendulum [8]. In fact, a real simple pendulum is often constructed using a metal sphere suspended on a string. In order to treat this pendulum as ideal, the metal sphere must have a small radius and a large mass compared to the length and mass of the string from which it is suspended. When the aim is to determine the acceleration of gravity, what should the minimum length of the string and the minimum mass of the sphere be to treat a real pendulum as ideal? The purpose of this paper is to answer this question. We, therefore, compare the period of oscillation of an ideal pendulum with that of a more realistic pendulum composed of a rigid sphere and a rigid slender rod. We determine the relative error in calculating the acceleration of gravity if the simple pendulum period is used instead of the "real" pendulum period.

11.2 Theoretical background

Figure 11.1 depicts a "real" simple pendulum constructed using a rigid sphere of mass m and radius R and suspended by a rigid slender rod of mass m' and length l. From elementary rotational dynamics of rigid bodies, we can write

$$\tau_O = -Mg\sin\theta L = I_O\frac{d^2\theta}{dt^2}, \quad (11.1)$$

where θ is the angular displacement measured from the downward vertical, τ_O is the summation of torques and I_O the moment of inertia of the pendulum relative to the pivot point O, $M = m' + m$ is the total mass of the pendulum, and

$$L = \frac{m'l/2 + m(l+R)}{M}, \quad (11.2)$$

is the distance between the pivot point O and the center of mass (CM) of the pendulum. For small angular displacements, the approximation $\sin\theta \approx \theta$ is valid and hence:

$$\frac{d^2\theta}{dt^2} + \frac{MgL}{I_O}\theta = 0. \quad (11.3)$$

From Eq. (11.3) the period of the pendulum is

$$T = 2\pi\sqrt{\frac{I_O}{MgL}}, \quad (11.4)$$

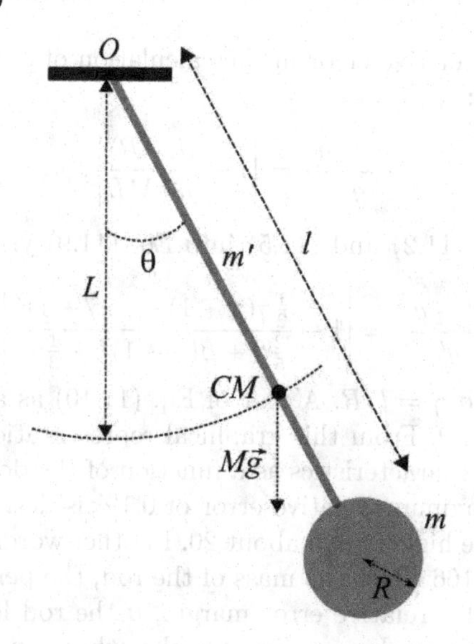

FIGURE 11.1
Schematic representation of a classroom pendulum.

where the moment of inertia of the pendulum relatively to the pivot point O is

$$I_O = \frac{2}{5}mR^2 + m(l+R)^2 + \frac{1}{3}m'l^2. \quad (11.5)$$

If the mass of the rod is negligible ($m' \ll m$), and the moment of inertia of the sphere can be neglected ($R^2 \ll l^2$), then $L = l + R$ and $I_O = mL^2$, and the period is given by the well-known simple pendulum period formula:

$$T = 2\pi\sqrt{\frac{L}{g}}. \quad (11.6)$$

11.3 The calculation of g

Equation (11.6) is commonly used to determine the acceleration of gravity by measuring the pendulum total length and period T:

$$g = \frac{4\pi^2}{T^2}(l+R). \quad (11.7)$$

However, according to Eq. (11.4), the correct way of calculating g should be:

$$g' = \frac{4\pi^2}{T^2}\frac{I_0}{ML}. \quad (11.8)$$

Let us calculate the relative error in the calculation of g if Eq. (11.7) is used instead of Eq. (11.8):

$$\frac{|\, g' - g\,|}{g'} = \left|1 - \frac{(l+R)}{I_0/ML}\right|. \tag{11.9}$$

Substitution of Eqs. (11.2) and (11.5) into Eq. (11.9) yields for the relative error:

$$\frac{|\, g' - g\,|}{g'} = \left|1 - \frac{\frac{1}{2}\gamma(\gamma+1) + \beta(\gamma+1)^2}{\frac{2}{5}\beta + \beta(\gamma+1)^2 + \frac{1}{3}\gamma^2}\right|, \tag{11.10}$$

where $\beta = m/m'$ and $\gamma = l/R$. A plot of Eq. (11.10) as a function of γ and β is shown in Fig. 11.2. From this graphical representation, it is possible to select the pendulum characteristics as a function of the desired relative error. For example, if a maximum relative error of 0.1% is desired and β is higher than 166, γ should be higher than about 20. In other words, if the mass of the sphere is more than 166 times the mass of the rod, the pendulum will behave as ideal, within a 0.1% relative error margin, if the rod length is more than 20 times larger than the sphere radius. On the other hand, if β is lower than 166, it is still possible to consider the pendulum ideal but only for a limited range of values. For example, if $\beta = 83$, the rod length must be 10 to 18 times larger than the sphere radius. Otherwise, the relative error will be higher than 0.1%.

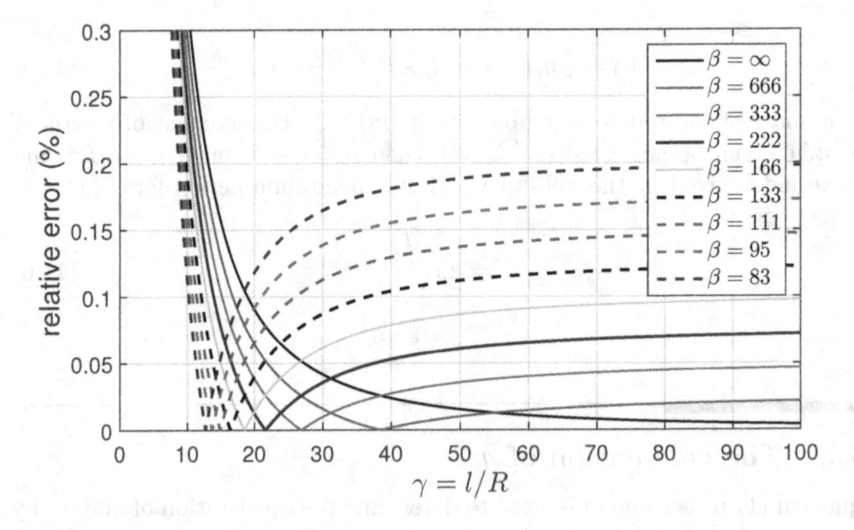

FIGURE 11.2
Relative error as a function of $\beta = m/m'$ and $\gamma = l/R$.

Interestingly, Fig. 11.2 shows that for each β, there is a value of γ for which the relative error is zero. We can find this optimal value by equating Eq. (11.10) to 0 and solving it for γ:

$$\gamma = -1.5 + \sqrt{\frac{9}{4} + \frac{12\beta}{5}}. \tag{11.11}$$

Equation (11.11) is plotted in Fig. 11.3 as a function of β. For example, for β=100, the relative error will be 0 if γ is approximately equal to 14. This optimal value is explained by the existence of two sources of error that acts in opposite ways: the radius of the sphere, which tends to increase the pendulum period, and the mass of the rod, which tends to decrease the pendulum period. As a result, the optimal value for which the relative error is null is achieved when the sphere's radius exactly cancels the effect of the rod mass.

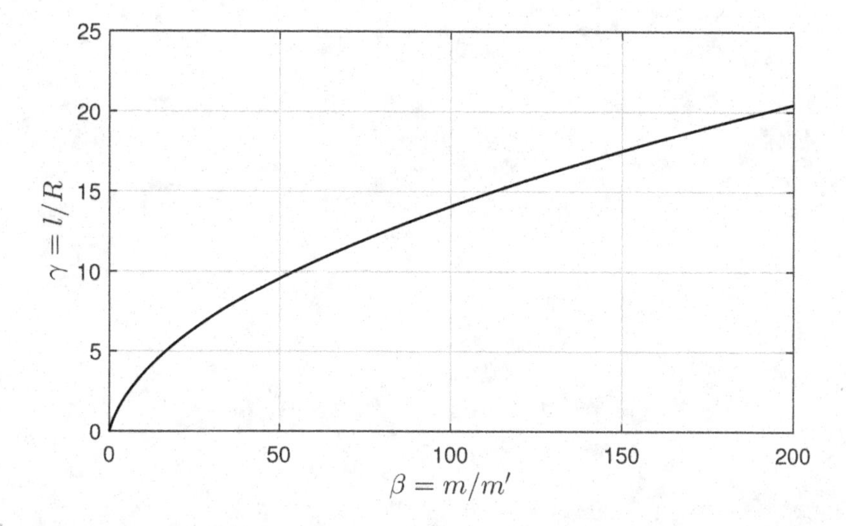

FIGURE 11.3
Plot of Eq. (11.11) showing the values of γ and β for which the relative error is null.

11.4 Conclusions

We have determined the characteristics of a classroom pendulum, constructed using a rigid sphere and a rigid slender rod, to be considered ideal within a certain relative error. For example, to have a maximum relative error of 0.1%, the rod should be more than 20 times longer than the radius of the sphere, if the mass of the sphere is more than 166 times the mass of the rod; the rod should be between 12 to 18 times larger than the radius of the sphere if the mass of the sphere is less than 166 times the rod mass. In addition, the relative error may be exactly zero if specific values of the ratios between the masses of the sphere and the rod and the length of the rod and the sphere radius are chosen.

References

[1] R. A. Serway and J. W. Jewett, *Physics for Scientists and Engineers with Modern Physics*, 8th ed. (Brooks/Cole, Belmont, CA, 2010), pp. 448–451.

[2] R. A. Nelson and M. C. Olsson, "The pendulum-Rich physics from a simple system," Am. J. Phys. **54**, 112–121 (1986).

[3] G. L. Baker and J. A. Blackburn, *The Pendulum: A Case Study in Physics* (Oxford University Press, New York, NY, 2005).

[4] K. Johannessen, "An approximate solution to the equation of motion for large-angle oscillations of the simple pendulum with initial velocity," Eur. J. Phys. **31**, 511–518 (2010).

[5] R. B. Kidd and S. L. Fogg, "A simple formula for the large-angle pendulum period" The Physics Teacher **40**, 81–83 (2002).

[6] F. M. S. Lima and P. Arun, "An accurate formula for the period of a simple pendulum oscillating beyond the small angle regime," Am. J. Phys. **74**, 892–895 (2006).

[7] C. G. Carvalhaes and P. Suppes, "Approximations for the period of the simple pendulum based on the arithmetic-geometric mean," Am. J. Phys. **76**, 1150–1154 (2008).

[8] T. H. Richardson and S. A. Brittle, "Physical pendulum experiments to enhance the understanding of moments of inertia and simple harmonic motion," Phys. Educ. **47**, 537–544 (2012).

12

Experiments with a Falling Rod

V. Oliveira

CONTENTS

We study the motion of a uniform thin rod released from rest, with the bottom end initially in contact with a horizontal surface. Our focus here is the motion of the bottom end as the rod falls. For small angles of release with respect to the horizontal, the rod falls without the bottom end slipping. For larger angles, the slipping direction depends on the static friction coefficient between the rod and the horizontal surface. Small friction coefficients cause the end to slip initially in one direction and then in the other, while for high coefficients, the end slips in one direction only. For intermediate values, depending on the angle of release, both situations can occur. We find the initial slipping direction to depend on a relation between the angle at which the rod slips and a critical angle at which the frictional force vanishes. Comparison between experimental data and numerical simulations shows good agreement.

Reproduced from V. Oliveira, "Experiments with a falling rod," American Journal of Physics **84**, 113–117 (2016), https://doi.org/10.1119/1.4934950, with the permission of the American Association of Physics Teachers.

DOI: 10.1201/9781003187103-12

12.1 Introduction

The plane motion of rigid bodies is an essential topic in introductory physics courses for scientists and engineers. The motion of rods and other elongated objects fallen on horizontal surfaces, with one end initially in contact with the surface, is a common example used in many engineering textbooks to illustrate this topic [1–3]. However, most of the problems presented in these textbooks tend to restrict the motion to specific conditions. For example, a typical problem involves finding the angular acceleration of the object immediately after its release. Such problems, although useful, constitute only a small part of the pedagogical richness offered by such a system. For instance, straightforward and interesting questions, such as at what angle or direction will the contact end slip, are generally not addressed.

The fall of rods and other elongated objects such as pencils, released with one end initially on a rough horizontal surface, has been studied theoretically by Turner and Pratt [4] and Cross [5]. From the relevant equations of motion, these authors show that the end initially in contact with the surface may slide either backward or forward, depending on the initial angle of inclination θ_0 and the coefficient of static friction μ_s. Thus, for example, if the rod is released from a stationary vertical position ($\theta_0 \approx 90°$) and $\mu_s < 0.371$, the rod slides backward at an angle greater than about 54.9°, while for $\mu_s > 0.371$ the rod slides forward for an initial angle between about 38.8° and 19.5°. Moreover, if $\mu_s < 0.371$, the rod slides initially backward for a while and then forward.

In the present paper, we study the fall of a rod from an experimental point of view. For this purpose, videos of experiments performed with a thin aluminum rod released on different surfaces are recorded with a high-speed camera and analyzed using a video analysis program. We find reasonably good agreement between experimental data and simulations obtained from the equations of motion.

12.2 Theoretical background

Figure 12.1 shows the free-body of a uniform rigid thin rod of mass m and length L released from rest at an angle θ_0 ($0° < \theta_0 < 90°$). Let point G be the rod's center-of-mass, point O be the contact point between the rod and the flat surface, hereafter designated as the "bottom end" of the rod, and θ the angle between the rod and the horizontal surface at some time t. For a uniform density rod, the distance between points G and 0 is $L/2$, and the moment of inertia of the rod about point O is $\frac{1}{3}mL^2$.

From the relevant equations of motion of the rod [4, 5], it can be shown that for non-slipping conditions the angular acceleration α of the rod is given

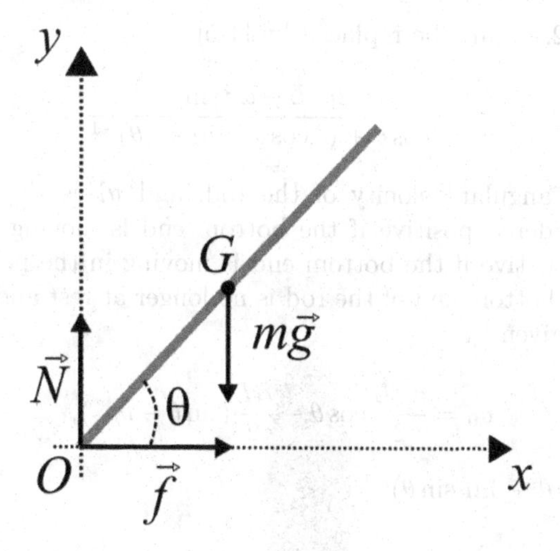

FIGURE 12.1
Free body diagram of a rod of mass m falling on a horizontal surface.

by

$$\alpha = \frac{3g}{2L}\cos\theta, \tag{12.1}$$

where α is considered positive if clockwise. Similarly, the frictional f and the normal N forces are given by

$$f = \frac{mg}{4}(9\cos\theta\sin\theta - 6\sin\theta_0\cos\theta), \tag{12.2}$$

and

$$N = \frac{mg}{4}(1 + 9\sin^2\theta - 6\sin\theta_0\sin\theta). \tag{12.3}$$

Equations (12.2) and (12.3) show that $f = 0$ at the critical angle $\theta_c = \sin^{-1}(\frac{2}{3}\sin\theta_0)$, while it is negative for $\theta < \theta_c$ and positive for $\theta > \theta_c$. The normal force is always positive except for $\theta_0 >= 90°$ and $\theta = \sin^{-1}(\frac{1}{3})$ when it is zero. In this particular case, the rod must slip before it reaches $\theta = \sin^{-1}(\frac{1}{3})$ no matter how large the static friction coefficient is because $|f/N|$ becomes arbitrarily large as this angle is approached. If slipping occurs at the angle $\theta = \theta_s$, then the static friction coefficient can be obtained from the condition of maximum static friction, and the definition that $\mu_s = f(\theta_s)/N(\theta_s)$, giving

$$\mu_s = \frac{|9\cos\theta_s\sin\theta_s - 6\sin\theta_0\cos\theta_s|}{1 + 9\sin^2\theta_s - 6\sin\theta_0\sin\theta_s}. \tag{12.4}$$

Equations (12.1), (12.2), and (12.3) are valid only as long as the static frictional force is sufficiently strong to prevent the rod from slipping. If the rod slips Eq. (12.1) must be replaced by [4,5]

$$\alpha = \frac{2g/L - \omega^2 \sin\theta}{\cos\theta + (3\cos\theta - \mu_k \sin\theta)^{-1}}, \tag{12.5}$$

where ω is the angular velocity of the rod, and μ_k is the kinetic friction coefficient, considered positive if the bottom end is moving in the negative direction and negative if the bottom end is moving in the positive direction. In addition, the bottom end of the rod is no longer at rest and moves with an acceleration a_0 given by

$$a_0 = \frac{\omega^2 L}{2}\cos\theta - \frac{\alpha L}{2}(\sin\theta - b\mu_k), \tag{12.6}$$

where $b = (3\cos\theta - 3\mu_k \sin\theta)^{-1}$.

12.3 Experiments and video analysis

In this section, we report on experiments performed with an aluminum cylindrical rod of mass m =55.4 g, length L =25.4 cm, and radius R=1.5 mm. The rod was released with the bottom end in contact with the surface of a steel sheet, a marble stone plate, and a mouse pad. A bubble level was used to make sure the surfaces were horizontal. All experiments were recorded with a high-speed video camera operating at 500 frames/s and analyzed using video analysis software. Prior to analysis, the videos were spatially calibrated using a ruler (present in the videos) and an origin assigned to the initial position of the bottom end of the rod. Both the angle θ and the horizontal position of the bottom end of the rod (x_0) were measured as a function of time for different angles of release. Before the main experiments, the static friction coefficient between the rod and each of the flat surfaces was calculated using Eq. (12.4). For this purpose, the slipping angle was determined from videos of the rod released from a nearly vertical position. Results are shown in Table 12.1. Note that the experimental error associated with some of these values is quite high because of the difficulty in determining the exact moment of slipping from the videos. Finally, three angles of release in the range 70°–80°, 20°–30°, and 5°–10° were selected in order to achieve $\theta_0 > \theta_s$, $\theta_0 < \theta_s$, and $\theta_0 \ll \theta_s$ for all surfaces.

12.3.1 Rod released on a steel surface

Typical video snapshots of the rod falling off the surface of a steel surface are displayed in Fig. 12.2 for θ_0 = 75.8°. Figure 12.3 shows the position of the bottom end of the rod as a function of the falling angle θ for initial angles

TABLE 12.1
Angle of release θ_0, slipping angle θ_s, and friction coefficient μ_s.

Surface	θ_0 (deg)	θ_s(deg)	μ_s
Cloth	85.0±0.2	33.5±1.0	1.9±0.3
Marble	85.5±0.2	36.2±1.0	0.9±0.3
Steel	86.0±0.2	67.6±0.5	0.27±0.01

of 75.8°, 26.0°, and 5.6°. For $\theta_0 = 75.8°$, three regimes can be distinguished. From the moment of release (Fig. 12.2a) until $\theta = 70.0°$ (Fig. 12.2b), the bottom end remained at rest; hence the rod falls without slipping. Beginning at θ=70.0°, the rod begins to slip backward, and the bottom end moves to the left. At $\theta = 19.5°$ (Fig. 12.2c), the bottom end momentarily comes to rest as it changes direction. The rod then slips forward as the bottom end moves to the right until the end of the fall (Fig. 12.2d). When the rod is released from 26.0°, similar behavior is observed as the bottom end moves first in the negative direction and then in the positive direction. However, in this case, the rod begins to slip the moment it is released. Finally, for $\theta_0 = 5.6°$, there is no slipping as the bottom end remains at rest during the entire fall.

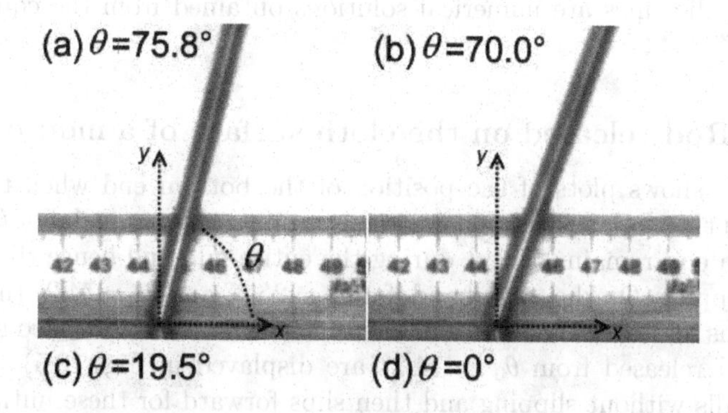

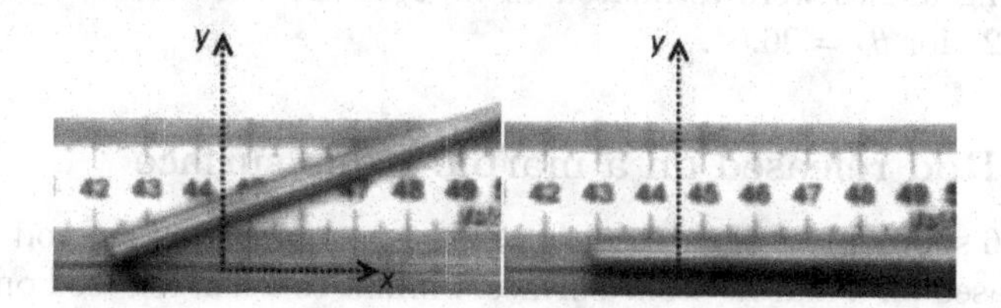

FIGURE 12.2
Video snapshots of a rod released on a steel surface from an initial angle of 75.8°.

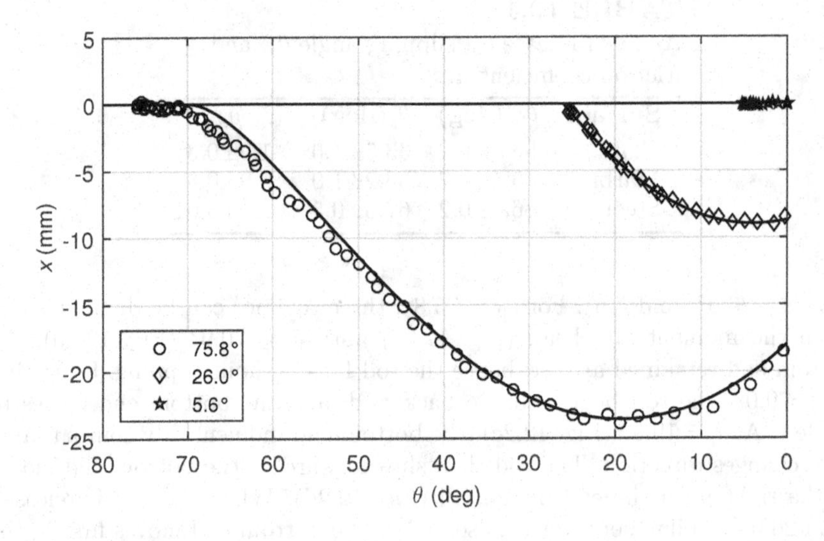

FIGURE 12.3
Horizontal displacement of the bottom end of the rod as a function of θ when the rod is released on a steel surface from initial angles of 75.8°, 26.0°, and 5.6°. The solid lines are numerical solutions obtained from the equations of motion.

12.3.2 Rod released on the cloth surface of a mouse pad

Figure 12.4 shows plots of the position of the bottom end when the rod is released on the cloth surface of a common computer mouse pad. For $\theta_0 = 9.8°$, the bottom end remains at rest during the entire fall, and hence the rod falls without slipping. On the other hand, for $\theta_0 = 26.6°$ and $\theta_0 = 74.2°$ the bottom end remains at rest for a while and then moves to the right (video snapshots for the rod released from $\theta_0 = 74.2°$ are displayed in Fig. 12.5). Thus the rod first falls without slipping and then slips forward for these initial angles. The slipping angles were estimated to be $\theta_s = 31 \pm 1°$ for $\theta_0 = 74.2°$ and $\theta_s = 9° \pm 2°$ for $\theta_0 = 26.6°$.

12.3.3 Rod released on a marble stone surface

Figure 12.6 shows plots of the position of the bottom end of the rod when the rod is released on a marble stone surface. Similar to when released on a mouse pad, when the rod is released from $\theta_0 = 77.0°$, the bottom end remains at rest initially and then moves in the positive direction. However, for $\theta_0 = 27.1°$, a completely different behavior is observed. Here, the bottom end moves in the negative direction and then moves in the positive direction near the end of the fall. This behavior is similar to the rod falling on the steel surface. For these cases, the slipping angles were estimated to be $\theta_s = 35 \pm 1°$ for $\theta_0 = 77.0°$

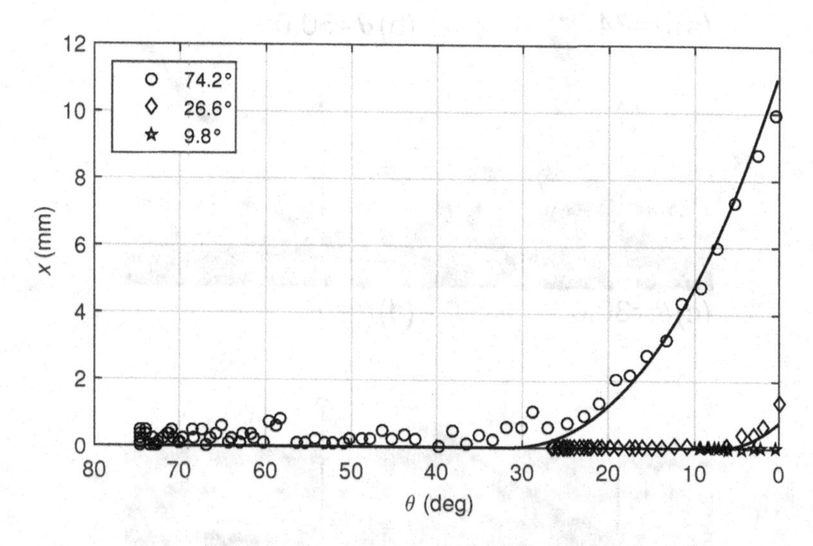

FIGURE 12.4
Horizontal displacement of the bottom end of the rod as a function of θ when the rod is released on the cloth surface of a computer mouse pad from initial angles of 74.2°, 26.6°, and 9.8°. The solid lines are numerical solutions obtained from the equations of motion.

and $\theta_s = 26.5 \pm 0.3°$ for $\theta_0 = 27.1°$. Finally, when the rod is released from $\theta_0 = 6.4°$, the bottom end does not move at all, and the rod falls without slipping (video snapshots for the rod released from $\theta_0 = 76.9°$ are displayed in Fig. 12.7).

12.4 Comparison to theory

The initial slipping behavior of the rod can be predicted theoretically by comparing θ_c and θ_s: for $\theta_s < \theta_0$ the frictional force is negative at the moment of slipping, and hence the rod slips forward; for $\theta_s > \theta_0$ the frictional force is positive, and the rod slips backward. However, because θ_c and θ_s are related to θ_0 and μ_s via $\theta_c = \sin^{-1}(\frac{2}{3}\sin\theta_0)$ and Eq. (12.4), the initial slipping behavior can also be predicted from μ_s and θ_0. Figure 12.8 depicts a phase diagram in the μ_s-θ_0 plane that gives the different behaviors that can occur. For $\mu_s > 0.75$, we always have $\theta_s < \theta_c$, and the rod does not slip backward. On the other hand, for $\mu_s < 0.75$ we may have $\theta_s < \theta_c$ or $\theta_s > \theta_c$ depending on θ_0, and hence the rod may slip in either direction (or remain still). Notice that slipping

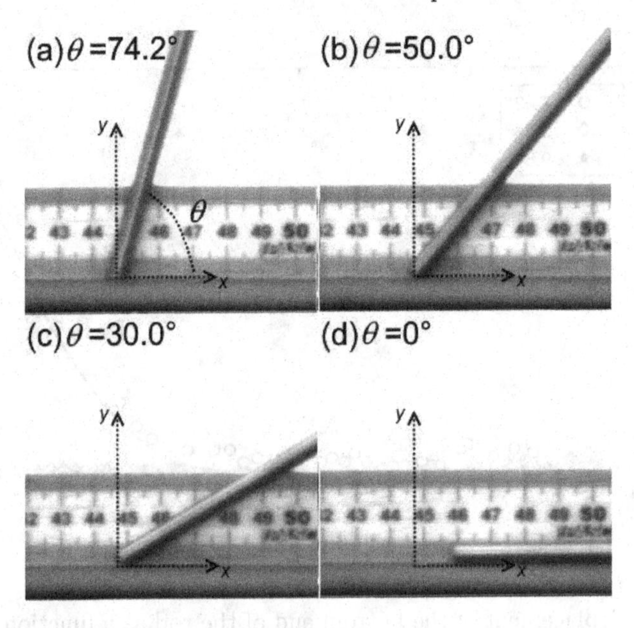

FIGURE 12.5
Video snapshots of the rod falling on the cloth surface of a mouse pad released from an initial angle of 74.2°.

can always be prevented by making $\theta_s \leq 0$, which can be accomplished by making μ_s very large or by making θ_0 very small.

In order to compare the experimental results with theoretical predictions, the horizontal position of the bottom end of the rod was determined numerically, solving the equations of motion of the rod. For this purpose, the angular acceleration is first determined using Eq. (12.1) if falling is without slipping, or Eq. (12.5) for falling and slipping. The (new) values of ω and θ are then determined numerically from α using a finite-difference scheme (time step of 0.001 s). Next, if the rod is slipping, a_0 is computed from Eq. (12.6) using the values of α, ω, and θ previously calculated; otherwise, a_0 is taken to be zero. Finally, the position (x_0) and speed (v_0) of the bottom end of the rod are determined numerically from a_0 using the same finite-difference method (time steps of 0.001 s). The static friction coefficients were taken from Table 12.1, while the kinetic friction coefficients were chosen to optimize the fit to the data. The values used were $\mu_s = 1.6$ and $\mu_k = 0.8$ for cloth, $\mu_s = 0.7$ and $\mu_k = 0.33$ for marble, and $\mu_s = 0.26$ and $\mu_k = 0.185$ for steel. The numerical results are in reasonably good agreement with the experimental data, as shown in Figs. 12.3, 12.4, and 12.6.

Table 12.2 shows a comparison between the experimental and theoretical values of θ_s, along with the theoretical value of θ_c, for all of the experiments. In addition, the initial slipping directions observed from experiments and

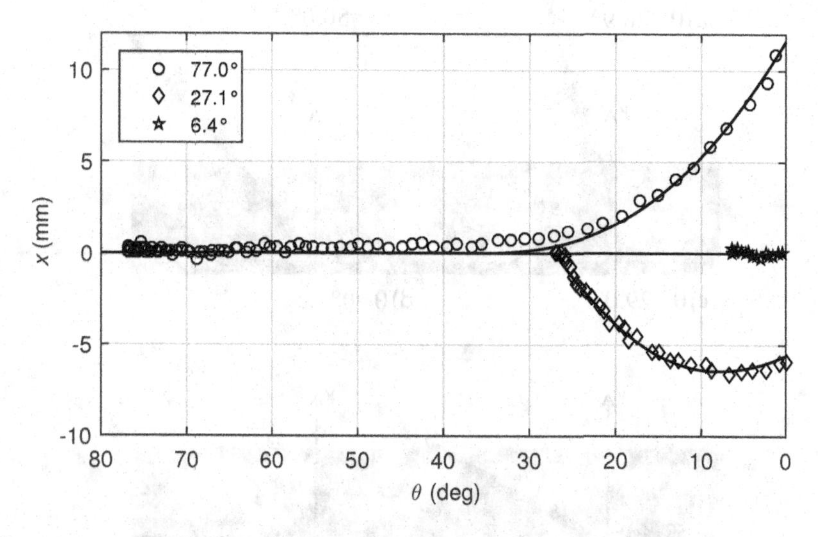

FIGURE 12.6
Horizontal displacement of the bottom end of the rod as a function of θ when the rod is released on a marble stone from initial angles of 6.4°, 27.1°, and 77.0°. The solid lines are numerical solutions obtained from the equations of motion.

predicted from Fig. 12.7 are also presented. The agreement between measured and predicted slipping angles θ_s is quite good, as all theoretical values (except one) fall within the margin of error in the experimental data. For the slipping direction, there is a discrepancy between theory and experiment for steel at the smallest initial angle. We believe this discrepancy is due to the difficulty in detecting very small movements of the bottom end of the rod from the videos. In addition, Table 12.2 shows that $\theta_s < \theta_c$, the rod slips forward, while for $\theta_s > \theta_c$ the rod slips backward, as expected from Eq. (12.2).

An intriguing aspect of this system is that for $\theta_s > \theta_c$, the initial backward slide gives way to a forward slide, whereas for $\theta_s < \theta_c$ the rod only slips forward. This behavior can be qualitatively understood with reference to Eq. (12.6). For $\theta_s > \theta_c$, the initial backward slide means $a_0 < 0$ and hence $\omega^2 \cos\theta < \alpha(\sin\theta - b\mu_k)$. Because this occurs near the beginning of the fall, ω is small, and θ is large, making the quantity $\omega^2 \cos\theta$ small. But as the fall progresses, ω increases and θ decreases, both of which act to increase $\omega^2 \cos\theta$. The fact that $\omega^2 \cos\theta$ increases faster than $\alpha(\sin\theta - b\mu_k)$ eventually leads to a positive a_0 and a forward slide. For $\theta_s < \theta_c$, the situation is different because the bottom end of the rod remains at rest for a longer period. As a result, when sliding occurs, ω is already large, and θ is small, resulting in a_0 being positive for the entire fall.

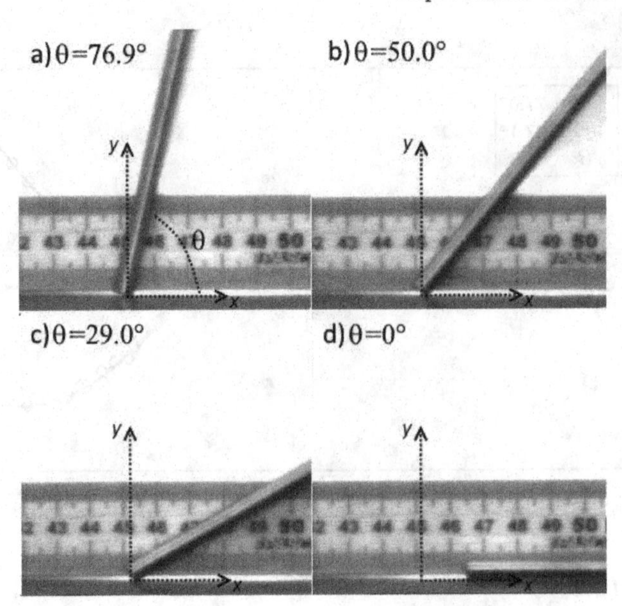

FIGURE 12.7
Video snapshots of the rod falling on the marble stone surface released from an initial angle of 76.9°.

12.5 Conclusions

Although the motion of a uniform thin rod released from rest on a horizontal surface is a well-known system, only theoretical descriptions are available in the literature. This paper has studied the rod's motion from an experimental point of view using a high-speed video camera. We studied the influence of different types of horizontal surfaces and multiple release angles. For very low release angles, the rod does not slip at all, while for higher angles, the rod may either slip forward or backward and then forward, depending on the type of horizontal surface. We also presented numerical simulations of the equations of motion of the falling rod that agreed quite well with the experimental data.

It is interesting to note that the present system and methodology can be easily used to determine static friction coefficients in the classroom. For example, if the rod is released from a near-vertical position, then the static friction coefficient μ_s can be determined from the slipping angle θ_s via Eq. (12.4). As a result, students can estimate the static friction coefficient of different surfaces by determining the corresponding slipping angles from videos. Note that, for this purpose, the rod should be released from a near-vertical position to guarantee that slipping occurs.

TABLE 12.2
The sliping angle θ, obtained from experiments and numerical simulations. The corresponding values of θ_0 and θ_c, and the initial slipping direction of the rod are also showed. Forward (fwd) means the bottom end of the rod is moving along the positive x-direction, backward (bwd) means it moves along the negative x-direction, and does not slip (dns) means the bottom end of the rod remains still.

		θ_s (deg)			Sliping direction	
Surface	θ_0**(deg)**	**exp.**	**theo.**	θ_c**(deg)**	**exp.**	**theo.**
cloth	74.2	31±1	31.8	39.9	fwd	fwd
$\mu_s = 1.9\pm0.3$	26.6	9±2	7.3	17.4	fwd	fwd
	9.8	−−	−−	6.5	dns	dns
marble	77.0	35±1	35.9	40.5	fwd	fwd
$\mu_s = 0.9\pm0.3$	27.1	27±1	27.1	17.7	bwd	bwd
	6.4	−−	−−	4.3	dns	dns
steel	75.8	70±1	69.9	40.3	bwd	bwd
$\mu_s = 0.27\pm0.01$	26.0	26±1	26.0	17.0	bwd	bwd
	5.6	−−	−−	4.3	dns	bwd

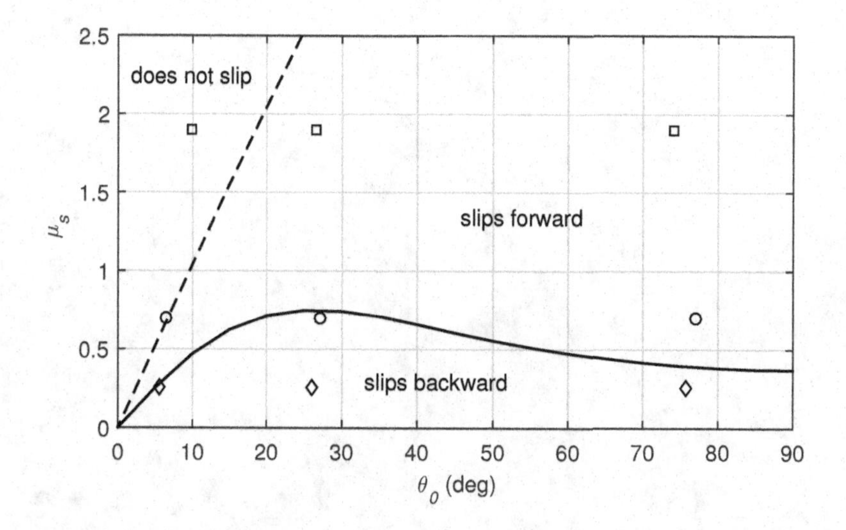

FIGURE 12.8
Static friction coefficients and initial slipping directions. If the value of μ_s falls below the solid line for a given θ_0, the rod will slip backward; if it falls between the solid and dashed lines, the rod will slip forward; if it falls above the dashed line, the rod does not slip at all. The symbols show the slipping directions expected for the angles of release used in the experiments for static friction coefficients of 1.9 for cloth, 0.7 for marble, and 0.26 for steel.

References

[1] J. L. Meriam and L. G. Kraige, *Engineering Mechanics: Dynamics* (John Wiley and Sons, Hoboken, NJ, 2017), p. 456.

[2] A. Pytel and J. Kiusalaas, *Engineering Mechanics: Dynamics*, 3rd ed. (Cengage Learning, Stamford, CT, 2009), p. 388.

[3] D. Halliday, R. Resnick, and J. Walker, *Fundamentals of Physics*, 8th ed. (John Wiley and Sons, Hoboken, NJ, 2007), problem 63, p. 270.

[4] R. Cross, "The fall and bounce of pencils and other elongated objects," Am. J. Phys. **74**, 26–30 (2006).

[5] L. Turner and J. L. Pratt, "Does a falling pencil levitate?," Quantum **8**, 22–25 (1998).

13

Oscillations of a Rectangular Plate

V. Oliveira

CONTENTS

We study the motion of a rectangular plate that oscillates around an axis perpendicular to its plane or parallel to its minor side. The pendulum is connected to a rotary motion sensor, and its angular position is measured as a function of time. Among others, we study the effect of the pendulum's initial angular displacement, mass, distance between the pivot point and the center of mass on the oscillation period, and the impact of different types of damping.

13.1 Introduction

The physical pendulum is part of most introductory physics courses covering rigid bodies' plane motion or oscillatory motion physics. For this reason, it has been widely studied in the literature [1–6]. From a practical point of view, physical pendulums are usually rectangular plates that oscillate around a fixed point. This system shows up many interesting situations and helps students understand key concepts such as the moment of inertia and torque. An oscillating rectangular plate may exhibit harmonic or anharmonic motion [4, 7, 8], and its motion can be damped by dry and viscous dissipative forces [2–4, 9–11]. In this context, this chapter aims to study the oscillatory motion of rectangular plates. The experimental data are compared with theoretical predictions obtained from analytical or approximate solutions of the plate's

DOI: 10.1201/9781003187103-13

motion equation. From a pedagogical perspective, this is very useful since it allows the validation of the models used or shows their limitations.

13.2 Experimental setup

Figure 13.1 shows a schematic representation of the physical pendulum used in this work. It is composed of a rectangular plate of sides $a = 5.0$ cm and $b = 35.3$ cm. The plate can swing around two axes passing by the pivot point O, one perpendicular to it (z-axis) and another parallel to its minor side (x-axis), and the distance L between the pendulum's center of mass and the pivot point can be adjusted. Four plates are used to study the effect of the pendulum's mass, two made of paperboard, with masses $m_1 = 6.6$ g (plate 1) and $m_2 = 14.3$ g (plate 2), and two made of wood, with masses $m_2 = 48.4$ g (plate 3) and 101.8 g (plate 4). The experimental data is acquired by fixing the plate to a rotary motion sensor Pasco PS-2120 and using the commercial software Pasco Capstone. First, the angular position is zeroed with the plate hanging vertically at rest. Next, the plate is rotated by hand up to the desired initial angle θ_0 and then released and left to oscillate.

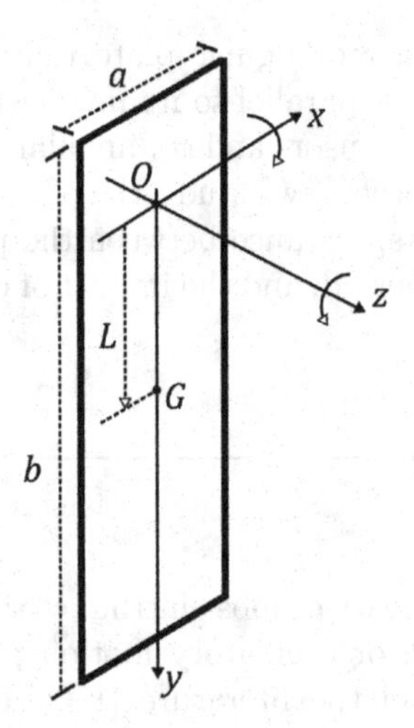

FIGURE 13.1
A schematic representation of the physical pendulum and each axis of rotation. L is the distance between the pivot point O and the center of mass G.

13.3 Results and Discussion

We first study the physical pendulum's behavior when small-amplitude oscillations are induced along the z-axis. For this configuration, we study the effect of L and θ_0 on the period of oscillation and the effect of the pendulum's mass on the degree of damping. Next, we study the behavior of the pendulum for oscillations along the x-axis. Here, we show that the damping mechanism depends on the mass of the pendulum.

13.3.1 Oscillations along the z-axis

Figure 13.2 depicts the angular displacement as a function of time for the four plates when released from an initial angle of about 0.2 rad. At first glance, the behavior of the plates seems to be strongly influenced by the mass. In all cases, the amplitude decreases with time, but the decay is more severe for small masses. The oscillation period is approximately constant, and its average value is about $T_{exp} = 0.952 \pm 0.003$ s.

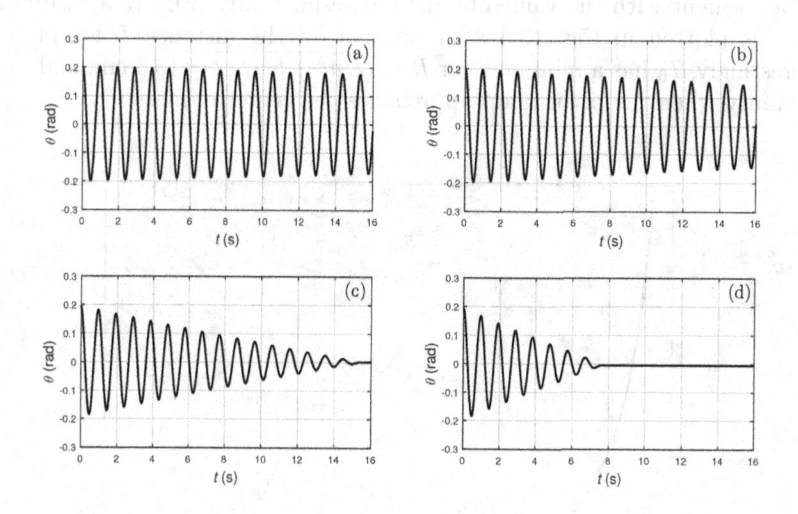

FIGURE 13.2
Angular displacement along the z-axis for plates (a) 4, (b) 3, (c) 2, and (d) 1, for $L = 15.65$ cm and $\theta_0 = 0.2$ rad.

If we neglect, for now, the damping that causes the amplitude's decay, the pendulum's equation of motion can be expressed as

$$I_O\ddot{\theta} = -mgL\sin\theta, \tag{13.1}$$

where θ is measured from the downward vertical, m is the pendulum's mass, g is the acceleration of gravity, $\ddot{\theta}$ is the angular acceleration, and $I_O = m(a^2 + b^2)/12 + mL^2$ is the moment of inertia of the pendulum relative to the z-axis. Assuming small angular displacements, we can use the approximation $\sin\theta \simeq \theta$ and write Eq. (13.1) in the simple form

$$\ddot{\theta} + \omega_0^2\theta = 0, \tag{13.2}$$

where $\omega_0 = \sqrt{mgL/I_O}$. The solution of this differential equation satisfying the initial conditions $\theta(t=0) = \theta_0$ and $\dot{\theta}(t=0) = 0$ is

$$\theta(t) = \theta_0 \cos\omega_0 t. \tag{13.3}$$

In this regime, the plate oscillates with a period $T_0 = 2\pi/\omega_0$ and, hence,

$$T_0 = 2\pi\sqrt{\frac{(a^2+b^2)/12 + L^2}{gL}}. \tag{13.4}$$

Substituting in this equation $L = (15.65\pm0.05)$ cm, $g = (9.80\pm0.01)$ m/s^2, $a = (5.00 \pm 0.05)$ cm, and $b = (35.30 \pm 0.05)$ cm, we obtain $T_0 = (0.950 \pm 0.002)$ s in agreement with the value obtained experimentally (Fig. 13.3). Equation (13.4) is plotted in Fig. 13.3 as a function of the distance L for plate 3. Interestingly, T_0 has a minimum for $L = \sqrt{(a^2+b^2)/12} = 10.3$ cm, and tends to infinity when the pivot point approaches the center of mass.

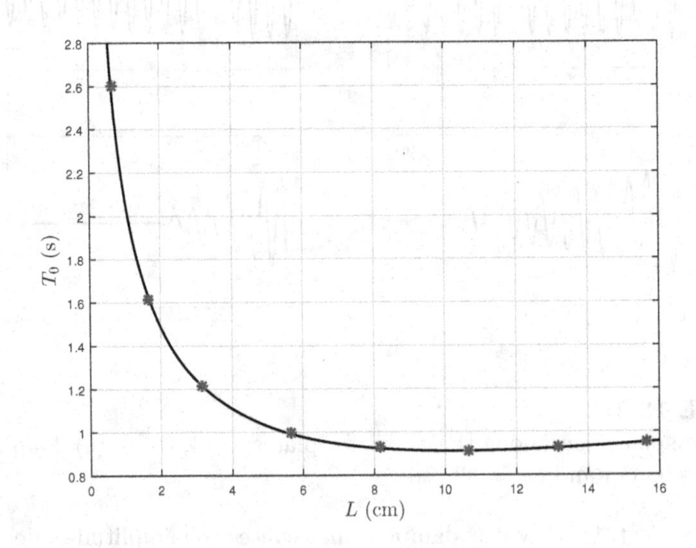

FIGURE 13.3
Period as a function of L. The asterisks are experimental points acquired with plate 3. The solid line shows the period calculated by Eq. (13.4).

If the small amplitude approximation cannot be used, the pendulum's period must be found solving Eq. (13.1). An expression for the exact period T may be derived from energy considerations [12,13]. If we choose the initial conditions $\theta(t=0)=\theta_0$ and $\dot{\theta}(t=0)=0$, we may write

$$mgL(1-\cos\theta_0)=\frac{1}{2}I_O\left(\frac{d\theta}{dt}\right)^2+mgL(1-\cos\theta). \tag{13.5}$$

from which, using the trigonometric identity $\cos\theta = 1-2\sin^2(\theta/2)$, we get

$$dt=\frac{1}{2\omega_0}[\sin^2(\theta/2)-\sin^2(\theta/2)]^{-\frac{1}{2}}d\theta. \tag{13.6}$$

Since the motion is symmetric, integrating Eq. (13.6) from $\theta=0$ to $\theta=\theta_0$ yields $T/4$ and hence:

$$T=\frac{T_o}{\pi}\int_0^{\theta_0}[\sin^2(\theta/2)-\sin^2(\theta/2)]^{-\frac{1}{2}}d\theta. \tag{13.7}$$

where T_0 is given by Eq. (13.4). Finally, if we make the substitution $\sin\phi = \sin(\theta/2)/\sin(\theta_0/2)$, Eq. (13.7) becomes:

$$T=\frac{2T_0}{\pi}\int_0^{\pi/2}\frac{d\phi}{\sqrt{1-\sin^2(\theta_0/2)\sin^2\phi}}. \tag{13.8}$$

The definite integral in Eq. (13.8) is known as a complete integral of the first kind, and it can be numerically evaluated for a given value of θ_0. Note that for small oscillations ($\theta_0 \to 0$), this integral equals $\pi/2$ and hence $T=T_0$. On the other hand, as θ_0 increases, the value of T diverges significantly from T_0. Figure 13.4 compares the period of plate 3 for angular displacements up to about 3 rad, with the ones predicted by Eqs. (13.4) and (13.8). Equation (13.4) underestimates the exact period for large amplitudes, but the difference is unnoticeable below 0.2 rad. On the other hand, the values calculated using Eq. (13.8) fit the experimental data perfectly until about 1.5 rad. Above this value, a slight deviation is observed, probably caused by non-negligible air resistance at large oscillations [7,14].

Figures 13.2c and d show that the amplitude decay cannot be ignored for plates 1 and 2. A plot of the amplitude as a function of the number of half-oscillations (Fig. 13.5) reveals that the decay is linear, which is typical of damping induced by friction in the bearings [3,10]. Note that the plates' surface subjected to air resistance is small for this configuration, so losses due to air drag are negligible.

The effect of friction in the bearings is usually modeled assuming a constant frictional torque τ_f [3,10]. For small angular displacements, the equation of motion becomes

$$\ddot{\theta}+\omega_0^2\theta=-a_1 sgn(\dot{\theta}), \tag{13.9}$$

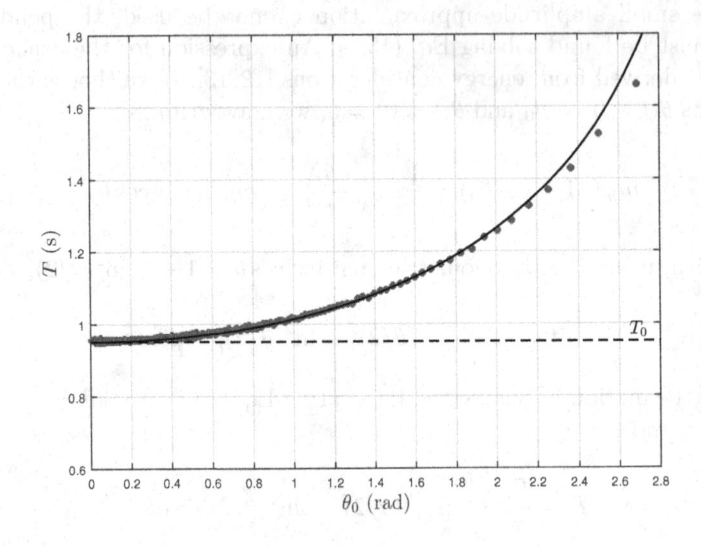

FIGURE 13.4
Period as a function of the amplitude. The asterisks are experimental data acquired with plate 3 suspended from L = 15.65 cm. The solid line is the period calculated using Eq. (13.8). The dashed line represents T_0.

where $a_1 = \tau_f/I_0$, $\dot{\theta}$ is the angular velocity, and *sgn* is the signum function, introduced here, so the frictional torque is always opposed to the angular velocity. The solution of Eq. (13.9) is [3]

$$\theta(t) = \left[\theta_0 - (2n+1)\frac{a_1}{\omega_0^2}\right]\cos(\omega_0 t) + (-1)^n\frac{a_1}{\omega_0^2}, \qquad (13.10)$$

for $n\pi/\omega_0 \leq t \leq (n+1)\pi/\omega_0$ and $n = 0, 1, 2, 3...$ is the number of half-oscillations. Note that the period remains $T_0 = 2\pi/\omega_0$ in this model. From Eq. (13.10) we can find an expression for the amplitude at the start of each half-oscillation:

$$\theta_n = \theta_0 - n\beta, \qquad (13.11)$$

where

$$\beta = \frac{2a_1}{\omega_0^2} = \frac{2\tau_f}{mgL} \qquad (13.12)$$

is the amplitude decay per half-oscillation. Equation (13.11) predicts a linear decrease of θ_n with n in agreement with the experimental data shown in Fig. 13.5. Interestingly, β is inversely proportional to the plate's mass and the distance L. Figure 13.6 shows the experimental values of β as a function of L for plates 1 to 3, determined from the linear fits to Eq. (13.11) shown in Fig. 13.5, and fits to Eq. (13.12).

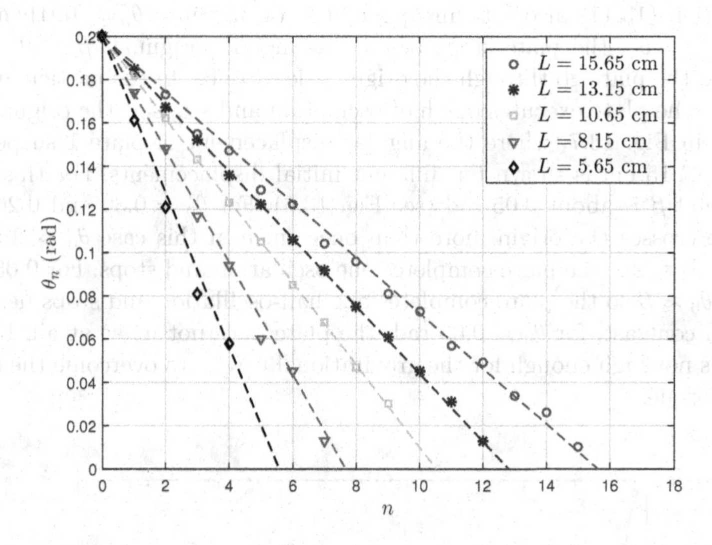

FIGURE 13.5
Maximum angular displacement of plate 1 along the z-axis as a function of the number of half-oscillations. The dashed lines are linear fits to the experimental points.

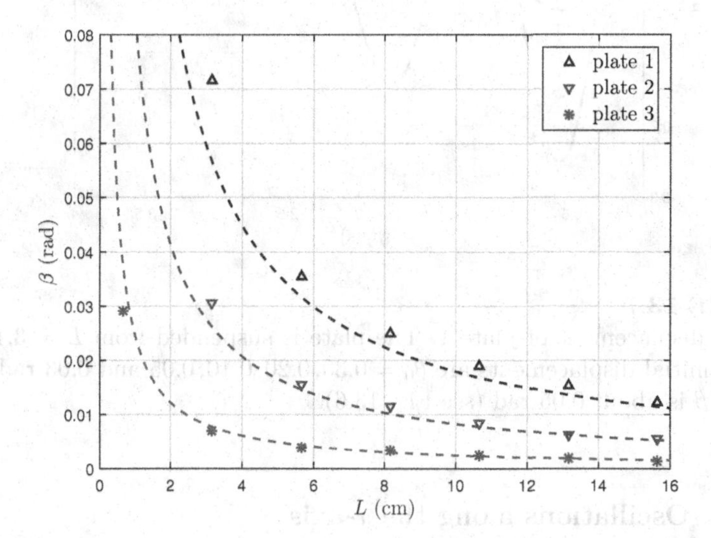

FIGURE 13.6
The parameter β as a function of L for plates 1 to 3. The dashed lines are fits to Eq. (13.12) with $\tau_f = 5.8 \times 10^{-5}$ Nm.

Equation (13.11) also sets an upper limit on n. Since $\theta_n \geq 0$, then $n \leq \theta_0/\beta$. If $\beta > \theta_0$, the plate stops before passing the origin. If $\beta < \theta_0$, then $n \geq 1$ and the plate go through the origin at least once. In the particular case of $\beta = \theta_0$, the plate executes one half-oscillation and stops at the origin. This is shown in Fig. 13.7, where the angular displacement of plate 1 suspended from $L = 3.15$ cm is shown for different initial displacements. For this case, the value of β is about 0.05 rad (see Fig. 13.6). For $\theta_0 = 0.30$ and 0.20 rad, the plate crosses the origin more than once since in this case $\theta_0 > 2\beta$. For 0.10 rad, $\theta_0 \approx 2\beta$, the plate completes one oscillation and stops. For 0.05 rad, we have $\theta_0 \approx \beta$ so the plate completes one half-oscillation and stops near the origin. In contrast, for $\theta_0 = 0.03$ rad, the plate does not move at all. In this case, θ_0 is not high enough for the gravitational torque to overcome the static friction torque.

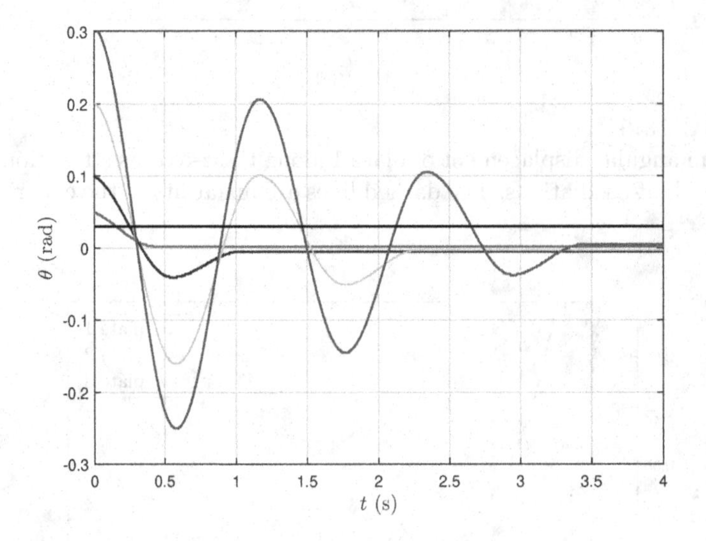

FIGURE 13.7
Angular displacement of plate 1. The plate is suspended from $L = 3.15$ cm and the initial displacements are $\theta_0 = 0.30, 0.20, 0.10, 0.05$ and 0.03 rad. The value of β is about 0.05 rad (see Fig. 13.6).

13.3.2 Oscillations along the x-axis

Figure 13.8 shows the angular displacement of the four plates along the x-axis. Note that the plate's surface subjected to air resistance is much higher than before, so it cannot be neglected in this configuration. Once again, the behavior of the plates is strongly influenced by their mass. The amplitude of the oscillations decays rapidly with time, but, in this case, the decrease is no longer linear. Semi-logarithmic plots of the amplitude vs. time are depicted in Fig. 13.9. Two regimes can be identified. The relation between $\ln(\theta_n)$ and time is linear for intermediates and large amplitudes, so the decay is essentially exponential for these amplitudes, and damping is mainly caused by air resistance. On the contrary, for low amplitudes, a downward curvature is observed, indicating a linear decay caused by dry friction at the bearings.

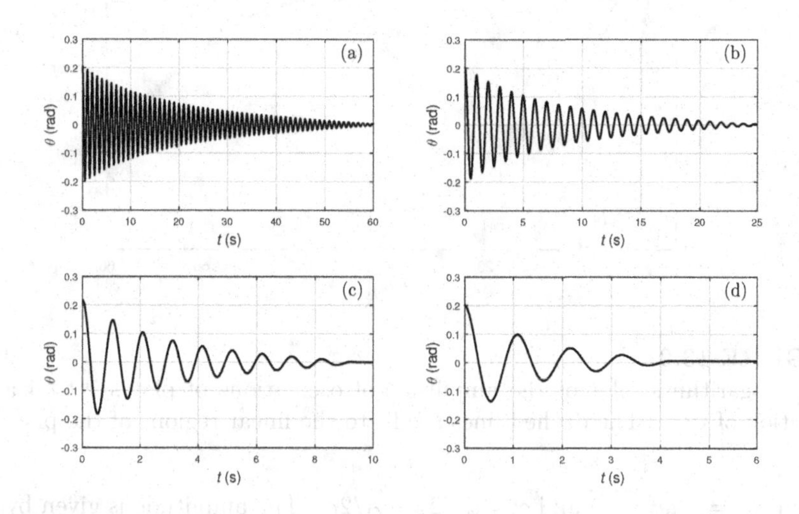

FIGURE 13.8
Angular displacement of plates (a) 1; (b) 2; (c) 3; and (d) 4, for $L = 20.65$ cm and $\theta_0 = 0.2$ rad.

To model the plate's behavior for intermediate and large angles, we assume that the drag force acting on the plate is perpendicular to it and has a magnitude proportional to its area A and velocity v [15]. Then, the total torque on the plate is given by [6],

$$\tau_d = cA(b^2/12 + L^2)\dot{\theta}. \tag{13.13}$$

where c is a constant. For small angular displacement, the equation of motion of the plate becomes

$$\ddot{\theta} + \omega_0^2\theta = -a_2\dot{\theta}, \tag{13.14}$$

where $a_2 = cA(b^2/12 + L^2)/I_0 = cA/m$. The solution of this equation satisfying the initial conditions $\theta(t = 0) = \theta_0$ and $\dot{\theta}(t = 0) = 0$ is

$$\theta = \theta_0 \exp(-\gamma t)[\cos(\omega_1 t) - \gamma/\omega_1 \sin(\omega_1 t)], \tag{13.15}$$

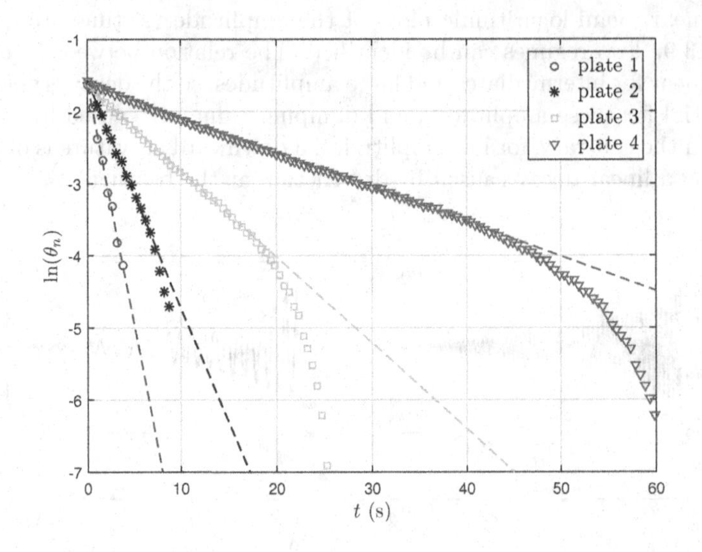

FIGURE 13.9
Semi logarithmic plot of the amplitude of oscillations of plates 1 to 4 as a function of time. The dashed lines are fit to the linear regions of the plots.

where $\omega_1 = \sqrt{\omega_0^2 - \gamma^2}$ and $\gamma = a_2/2 = cA/2m$. The amplitude is given by

$$\theta_n = \theta_0 \exp(-\gamma t). \tag{13.16}$$

The model predicts a linear decrease of $\ln(\theta_n)$ with time, following the experimental results for intermediate and large angles shown in Fig. 13.9. It also predicts that γ is inversely proportional to m but independent of L. Figure 13.10 compares a plot of γ as a function of m, using a fitted value of $c = 0.55$ kg s^{-1}m^{-2}, with experimental data calculated from the slope of the linear region of the plots of $\ln(\theta_n)$ shown in 13.9. A good agreement is found.

13.4 Conclusions

We study the oscillations of a rectangular plate. Depending on the plate's mass and axis of rotation, different types of motion can be identified. For example, if the plate oscillates around an axis perpendicular to its plane, its motion is damped due to friction at the bearings. For this configuration, the period decreases with decreasing amplitude and tends to a constant value given by Eq. (13.4) for small amplitudes. Except for very large amplitudes, the experimental values are well fitted by Eq. (13.8). On the other hand,

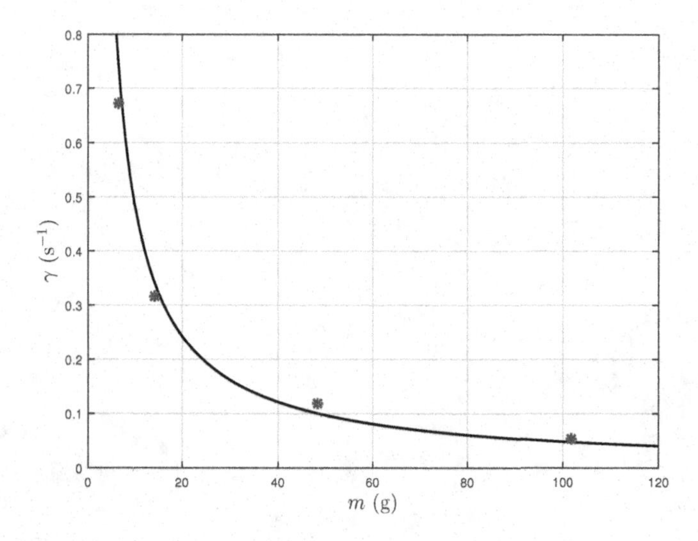

FIGURE 13.10
The parameter γ as a function of the mass. The asterisks are experimental values of γ calculated from the semi-logarithmic plots of the amplitude shown in Fig. 13.9. The solid line represents $cA/2m$ with $c = 0.55$ kg s^{-1}m^{-2}.

damping is well modeled, assuming a constant frictional torque. If the plate oscillates around an axis parallel to its plane, air drag cannot be neglected and becomes the primary damping mechanism. In this case, damping is well modeled, assuming a drag force proportional to the plate's angular velocity. In this case, the amplitude decay is exponential.

References

[1] I. M. Freeman, "Rectangular plate pendulum," Am. J. Phys. **22**, 157–158 (1954).

[2] P. Gluck, "Versatile physical pendulum," Phys. Teach. **42**, 226–230 (2004).

[3] J. C. Simbach and J. Priest, "Another look at a damped physical pendulum," Am. J. Phys. **73**, 1079–1080 (2005).

[4] J. C. Fernandes, P. J. Sebastião, L. N. Gonçalves, and A. Ferraz, "Study of large-angle anharmonic oscillations of a physical pendulum using an acceleration sensor," Eur. J. Phys. **38**, 045004 (2017).

[5] T. H. Richardson and S- A. Brittle, "Physical pendulum experiments to enhance the understanding of moments of inertia and simple harmonic motion," Phys. Educ. **47**, 537–544 (2012).

[6] V. Oliveira, "Oscillations of a plate made of A4 sheets of paper," Eur. J. Phys. **40**, 055005 (2019).

[7] F. M. S. Lima and P. Arun, "An accurate formula for the period of a simple pendulum oscillating beyond the small angle regime," Am. J. Phys. **74**, 892–895 (2006).

[8] R. B. Kidd and S. L.Fogg, "A simple formula for the large-angle pendulum period," Phys. Teach. **40**, 81–83 (2002).

[9] P. T. Squire, "Pendulum damping," Am. J. Phys. **54**, 984–991 (1986).

[10] L. F. C. Zonetti, A. S. S. Camargo, J. Sartori, D. F. de Sousa, and L. A. O. Nunes, "A demonstration of dry and viscous damping of an oscillating pendulum," Eur. J. Phys. **20**, 85–88 (1999).

[11] R. Hurtado-Velasco, Y. Villota-Narvaez, D. Florez, and H. Carrillo, "Video analysis-based estimation of bearing friction factors," Eur. J. Phys. **39**, 065807 (2018).

[12] S. T. Thornton and J. B. Marion, *Classical Dynamics of Particles and Systems*, 5th ed. (Brooks/Cole, Belmont, CA, 2004).

[13] S. D. Schery, "Design of an inexpensive pendulum for study of large-angle motion," Am. J. Phys. **44**, 666–670 (1976).

[14] R. A. Nelson and M. C. Olsson, "The pendulum–Rich physics from a simple system," Am. J. Phys. **54**, 112–121 (1986).

[15] J. A. Lock, "The physics of air resistance," Phys. Teach. **20**, 158–160 (1982).

14

Bullet Block Experiment: Angular Momentum Conservation and Kinetic Energy Dissipation

J. M. Tavares

CONTENTS

The analysis of a collision between a point particle (bullet) and a rigid body (block) that are under the action of gravity is revisited with the goal of explaining the results of an experiment conducted on a physics YouTube channel, very popular among students. In those videos, it is shown experimentally that: (i) blocks hit by bullets with the same velocity reach the same height, irrespectively of the position of the point of impact; (ii) blocks hit by bullets in impact points that are not aligned with the center of the block have an extra (rotational) energy after collision. In the present study, we show that these two results are a direct consequence of the application of basic laws of mechanics to a system of particles, regardless of the details of the interactions between the particles: (i) follows immediately from Newton's second and third laws; (ii) follows from the conservation of angular momentum and its relation with the dissipated energy.

DOI: 10.1201/9781003187103-14

14.1 Introduction

In the last few years, several physics channels have appeared on YouTube. These resources are becoming extremely popular among students since they often pose interesting questions, present experiments that usually are not performed at university and give intuitive and simple explanations of their results. One of the examples of the latter is a bullet block collision experiment presented by the channel *Veritasium* [1] in three consecutive videos [2–4] (that sum around 3.7 million views). In [2], it is announced that a bullet block experiment will be conducted. A bullet will be fired vertically upwards into a wooden block that, after the impact, will move with no constrains. In experiment 1, the bullet will be fired straight at the center of mass (CM) of the block, and, after impact, the block will reach a maximum height h_1. Experiment 2 will be performed with the same conditions, except that the bullet enters the block in a direction that does not contain the CM of the block. Viewers are asked to predict if the maximum height reached in the second experiment, h_2, is larger, smaller, or equal to h_1. A survey is conducted among several members of other YouTube physics channels: some predict that $h_1 = h_2$ (based on pure "intuition") and others that $h_1 > h_2$. The latter argue that since, in experiment 2, part of the kinetic energy of the bullet is transferred to the rotational kinetic energy of the block, the translational kinetic energy of the block after impact is smaller than in experiment 1 and thus, the maximum height reached has to be also smaller. The second video shows the experiment and its results: $h_1 = h_2$ (within experimental uncertainty), i.e. the block reaches always the same height when the bullet is fired vertically and upwards, irrespectively of the point where it hits the block. Viewers are invited to explain this apparently unexpected result. Finally, a third video explains that the result $h_1 = h_2$ is simply due to linear momentum conservation in the collision. The reason why, as a consequence, the dissipated kinetic energy is smaller in experiment 2, is left open, after an attempt (not verified experimentally) to explain it on the basis of the different paths of the bullet inside the block.

The immediate goal of this paper is to explain rigorously and in detail the origin of the two main results of the aforementioned experiments: $h_1 = h_2$ and the decrease of the dissipated kinetic energy from experiment 1 to experiment 2. We will put in evidence that the decrease in dissipated energy when the block rotates is a direct consequence of angular momentum conservation and does not need a justification based on the details of the block-bullet interactions during the collision. It turns out that the rigorous analysis of this problem involves a deep understanding of the concepts of linear momentum, angular momentum, inertia, and energy, applied to collisions. Therefore, the broader objective of this paper is to show that the popular experiments and discussions available on these new resources can be used in the classroom as a motivation for students to deepen their understanding of basic physical principles.

In the first section of this paper, we study a general plastic collision between a rigid body (block) and a point particle (bullet) in the presence of gravity; we establish general expressions for the height reached by the bullet+block system, the angular momentum, and the dissipated energy. In a second section, we apply this general approach to two particular types of blocks: a rod (with any mass distribution) and a homogeneous rectangular parallelepiped (the experimental situation in [2,3]). In the latter case, we discuss the possibility of verifying experimentally a possible relation between the dissipated energy and the length of the path traveled by the bullet inside the block during the collision.

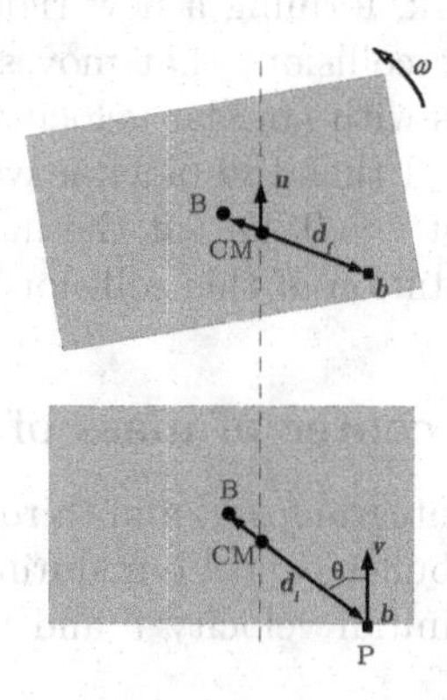

FIGURE 14.1
Schematic representation of the collision under analysis. A block, whose center of mass is point B, is hit by a bullet b with velocity $\vec{v}$ at point P at instant $t = 0$ (bottom panel). d_i is the distance between B and P, CM is the center of mass of the system bullet + block, and θ is the angle between the direction of the velocity and that of the vector that connects P to B. The collision lasts a time Δt. Immediately after the collision (top panel) the translation of the system is characterized by $\vec{u}$, the velocity of the CM, and the rotation of the system by $\vec{\omega}$, the angular velocity of the new rigid body formed by the bullet and the block. The distance between B and the bullet b is d_f. The dashed line represents the direction of gravity. In this representation $\vec{v}$ (and as a consequence $\vec{u}$) are vertical, but they could have any direction with a non-zero upwards component.

14.2 Plastic collision between a rigid body and a point particle

Let us consider a rigid body of mass M (called, for simplicity, the block, but which can have any shape or form) whose CM is point B, and a point particle of mass m (the bullet). Let us further consider that, in an inertial frame of reference (whose axes will be chosen later), the block is at rest and the bullet moves with velocity $\vec{v}$ when, at instant $t = 0$, it hits the block at a point P. Both the bullet and the block are under the action of gravity. The block is at rest before the collision because it is sustained by a support. During a subsequent time Δt, the bullet penetrates the block (collides with it) and, after that, remains inside the block, forming a new rigid body with mass $M + m$ (perfectly inelastic or plastic collision) that moves, without constrains, with CM velocity $\vec{u}(t)$ and rotates with angular velocity $\vec{\omega}(t)$ [1]. It is assumed that air resistance is negligible and that the block leaves the support (the action of the support becomes 0) at $t = 0$, i.e., at the instant when it is hit by the bullet. A schematic representation of this collision is depicted in Fig. 14.1.

14.2.1 Motion of the center of mass of the system

Newton's second law (in its integral form) and third law (the interaction forces between the block and the bullet cancel out during collision) establish a relation between the bullet's initial velocity $\vec{v}$ and the CM velocity $\vec{u}$ at time t,

$$(m + M)\vec{u}(t) = m\vec{v} + \int_0^t \vec{F}_{ext} dt, \tag{14.1}$$

where $\vec{F}_{ext}$ is the net external force acting on the system block+bullet. If $\vec{F}_{ext} = 0$, the system is isolated and its linear momentum is conserved. In the experiment under analysis, for $t > 0$, F_{ext} is simply the weight of the system and thus,

$$\vec{u}(t) = \frac{m}{m + M}\vec{v} + \vec{g}t \equiv \vec{u}(0) + \vec{g}t, \tag{14.2}$$

where $\vec{g}$ is the acceleration due to gravity, and $\vec{u}(0)$ is the initial velocity of the CM of the system. Therefore, during and after the collision, the CM of the system moves under the sole action of gravity. Noting that $\vec{v} \cdot \vec{g} < 0$, and integrating the vertical component of $\vec{u}(t)$, one obtains for the maximum height reached by the CM,

$$h_{CM,max} = h_{CM,0} + \frac{1}{2}\left(\frac{m}{m + M}\right)^2 \frac{v_g^2}{g}, \tag{14.3}$$

[1] Notice that the "real" case where the motion of the bullet causes a change in the mass distribution (e.g. by making a hole along its path inside the block) is not excluded from this analysis, provided that the mass M of the block remains constant.

where $h_{CM,0}$ is the height of the CM at the beginning of the collision and v_g is the vertical component of $\vec{v}$. Notice that $h_{CM,0}$ depends on m, M and on the initial heights of points B and P. One concludes that, for given $\vec{v}$, m and M, the maximum height reached by the CM of the system is independent of the position of the point of impact P, provided that the initial heights of P and B are the same (as happens in the experiments of [2–4]). This result is rather trivial: the motion of the CM of a system of particles always obeys $\vec{F}_{ext} = d\vec{p}_{CM}/dt$ (where $\vec{p}_{CM}$ is the linear momentum of the CM); since the external force is the weight, the *CM* of the system is, from the instant of impact on, simply moving as a projectile. For the same initial height and vertical component of the velocity, the CM will always reach the same maximum height. The details of the interactions between the bullet and the block, the spectacular collisions that can be appreciated in [3,4], the rotation of the system and the dissipation of energy, may produce, at the same time and paradoxically, a motivation and a distraction that combined can lead to hesitation (or even to error) in answer to the simple question posed in [2].

Usually, as is implicitly stated in [4], either the external forces during the collision (that are expected to be much smaller than internal forces) or its duration are neglected. In both cases ($\vec{F}_{ext} \approx 0$ or $\Delta t \approx 0$), these assumptions lead to linear momentum conservation, which, when assumed in Eq. (14.1), results also in Eq. (14.3). Here it is shown that, in this particular case, linear momentum conservation (and the hypothesis related to it) is not needed to obtain the final result. Finally, one should stress that the same height is reached (for different impact points P all at the same height) not by the CM of the block, point B, but by the CM of the system bullet+block. These heights are of course expected to be close to that reached by B if $m/M \ll 1$.

14.2.2 Conservation of angular momentum about the CM

The variation in time of the angular momentum of a system about its CM or about a fixed point of the body is equal to the total external torque on the system about the same point. Since the only external force acting on the system bullet+block at $t > 0$ is the weight, and since the torque of the weight about the CM is 0, the angular momentum about the CM, $\vec{L}_{CM}$, is conserved along the motion (both during and after collision). At time $t = 0$, $\vec{L}_{CM}$ can be calculated using the motion of the bullet,

$$\vec{L}_{CM} = m\vec{r}_{CM,P} \times \vec{v}, \tag{14.4}$$

where $\vec{r}_{i,j}$ is the vector that starts at point i and ends at point j (a notation we will use from now on). Choosing the z-axis to coincide with the direction of $\vec{L}_{CM}$, Eq. (14.4) becomes,

$$\vec{L}_{CM} = L_{CM}\hat{z} = \frac{mM}{m+M} v d_i \sin\theta \hat{z}, \tag{14.5}$$

where $\hat{z}$ is the unit vector associated to the z axis, θ is the angle between $\vec{v}$ and $\vec{r}_{CM,P}$, and $d_i = r_{CM,P}(m+M)/M$ is the distance between the center of mass of the block (point B) and the bullet (point P) at the beginning of the collision. Immediately after the collision, the bullet and the block form a rigid body and thus $\vec{L}_{CM}$ can be related to the angular velocity $\vec{\omega}$ through the tensorial relation,

$$\tilde{L}_{CM} = \tilde{I}\tilde{\omega}, \tag{14.6}$$

where $\tilde{I}$ is the inertia tensor of the bullet+block rigid body, and $\tilde{L}_{CM}$ and $\tilde{\omega}$ are representations of $\vec{L}_{CM}$ and $\vec{\omega}$ as column matrices. The inertia tensor $\tilde{I}$ is represented in a reference frame with origin at CM, with fixed axis (x, y, z) [2], the z-axis being parallel to $\vec{L}_{CM}$. Inverting Eq. (14.6), we can calculate the components of $\vec{\omega}$,

$$\begin{aligned} \omega_x &= \frac{I_{xy}I_{yz} - I_{yy}I_{xz}}{\det \tilde{I}} L_{CM} \equiv aL_{CM}, & (14.7) \\ \omega_y &= \frac{I_{xy}I_{xz} - I_{xx}I_{yz}}{\det \tilde{I}} L_{CM} \equiv bL_{CM}, & (14.8) \\ \omega_z &= cL_{CM}, & (14.9) \end{aligned}$$

with,

$$c = \frac{I_{xx}I_{yy} - I_{xy}^2}{\det \tilde{I}}, \tag{14.10}$$

where $I_{\alpha\beta}$ are the components of the inertia tensor, and $\det \tilde{I}$ is its determinant. If z is a principal axis of inertia of the system bullet+block, then $I_{xz} = I_{yz} = 0$ and, as a consequence, $a = b = 0$, $\omega_x = \omega_y = 0$, $\det I = I_{zz}(I_{xx}I_{yy} - I_{xy}^2)$ and $I_z\omega_z = L_{CM}$. Therefore, the system rotates around the z-axis with constant angular velocity at all instants $t > \Delta t$. In the general case, because the orientation of the rigid body relative to the fixed frame (x, y, z) changes in time, the angular velocity $\vec{\omega}$ is not constant: both a and b in Eqs. (14.7) and (14.8) (but not c in Eq. (14.9) as shown in the next section) vary in time, and, consequently, also ω_x and ω_y do. This can be seen in some of the images of [3,4], where the bullet+block system performs a rotation about an axis whose direction changes in time.

14.2.3 Rotational kinetic energy

The rotational kinetic energy of a rigid body is,

$$K_r = \frac{1}{2}\vec{L}_{CM} \cdot \vec{\omega}. \tag{14.11}$$

[2] If we wanted to study the rotational motion of the system in detail, then we would have to use a reference frame rotating with the body. Since this is not the goal of this work, this choice is simpler to prove the points that follow.

After the collision, the rotational kinetic energy of the rigid body bullet + block is, using Eq. (14.9),

$$K_r = \frac{1}{2} L_{CM}\omega_z = \frac{cL_{CM}^2}{2}. \tag{14.12}$$

Since the motion is torque-free, both $\vec{L}_{CM}$ and K_r are constant; therefore, even in the case of a general rotation, c is also constant. The rotational kinetic energy after the collision is thus the product of two terms: L_{CM}^2 is fixed by the bullet mass, the bullet velocity before the collision and by the position of the point of impact relative to the CM (see Eq. (14.5)); the constant c given by Eq. (14.10) is determined by the inertia tensor of the block, the position of the bullet inside the block after the collision, and by the orientation of the principal axis of inertia of the system relative to the direction of $\vec{L}_{CM}$.

14.2.4 Mechanical energy dissipated in the collision

The mechanical energy of the system before the collision (at $t = 0$) is the sum of the kinetic energy of the bullet with the gravitational potential energy of the system bullet+block,

$$E_i = \frac{1}{2}mv^2 + (m + M)gh_{CM,0} \tag{14.13}$$

The mechanical energy immediately after the collision (at instant $t = \Delta t$) is the sum of the kinetic energy (translational + rotational, the latter given by Eq. (14.11)) with the gravitational potential energy of the system,

$$E_f = \frac{1}{2}(m + M)u^2(\Delta t) + \frac{cL_{CM}^2}{2} + (m + M)gh_{CM}(\Delta t), \tag{14.14}$$

where $u(\Delta t)$ and $h_{CM}(\Delta t)$ are the velocity and height of the CM at $t = \Delta t$. Since the CM is moving with the acceleration of gravity, we have,

$$u^2(\Delta t) = u(0)^2 - 2g(h_{CM}(\Delta t) - h_{CM,0}). \tag{14.15}$$

Using Eqs. (14.2),(14.13),(14.14) and (14.15) one obtains for the dissipated mechanical energy,

$$E_d \equiv E_i - E_f = \frac{1}{2}mv^2\left(1 - \frac{m}{m + M}\right) - \frac{cL_{CM}^2}{2}. \tag{14.16}$$

This result shows immediately that the dissipated energy decreases when the angular momentum L_{CM} is increased. In the conditions of [2], where the bullet velocity is always vertical, and the linear momentum is the same in all experiments, the rotation of the system observed in the cases of off-center impacts is a consequence of angular momentum conservation. The energy of this rotation corresponds to energy that, in comparison with on-center impacts, has not been dissipated. Another interesting conclusion drawn from Eq. (14.16) is that the energy dissipated depends on the details of the collision (duration, characteristics of the materials, length of the path of the bullet inside the block) only through the value of c, since m, M, v, and L_{CM} are fixed before the collision.

The ratio of the mechanical energy associated with the CM after the collision to the initial mechanical energy E_i is (setting $h_{CM,0} = 0$),

$$\frac{E_{CM,f}}{E_i} = \frac{m}{m+M} \equiv \alpha. \tag{14.17}$$

This fraction depends only on the ratio m/M. If $m \ll M$ then $\alpha \approx m/M$ and only a tiny fraction of the initial energy is transformed into translational kinetic energy. The ratio of the rotational energy to E_i is (using Eqs. (14.11) and (14.5)),

$$\frac{K_r}{E_i} = cm(1-\alpha)^2 d_i^2 \sin^2\theta. \tag{14.18}$$

The fraction of initial energy that is transformed into rotational energy depends not only on α but on other parameters. In the next section, we analyze in detail some examples that show how the rotational kinetic energy of the system (and thus the dissipated energy) can be changed by changing the angular momentum and/or the inertia of the system.

It is worthwhile noticing that in the case of a collision between a rigid body that has a fixed point O and a bullet (that is often given as an application of angular momentum conservation in basic textbooks, e.g. [6]), a similar reasoning would apply. In this situation, linear momentum is not conserved (because of the external force acting on O that keeps this point fixed), but the relation between angular momentum and dissipated energy still stands. The larger the angular momentum about O is, the less energy is dissipated: when the bullet hits the block in a direction that contains O (i.e. when $L_O = 0$), all the mechanical energy is dissipated.

14.3 Dissipated energy and angular momentum conservation

In this section, we apply the general formalism to some examples, always assuming that the velocities of all points are parallel to a given plane (planar motion) and that the rotation is about a principal axis. We are particularly interested in exploring, in a more quantitative fashion, the relation between the dissipated energy, the conserved angular momentum, and the inertia of the system.

The vectors $\vec{r}_{B,P}$ and $\vec{v}$ define a plane xy that is perpendicular to a principal axis of rotation z of the block. It is assumed that, during the collision, the velocities of the center of mass of the block and of the bullet remain in the plane xy, and that, although the mass distribution of the block may be changing owing to the action of the bullet traveling through it, the z-axis is always a principal axis of inertia of the block.

The constant c in Eq. (14.10) is,

$$c = \frac{1}{I_{zz}}, \tag{14.19}$$

since it is assumed that z is a principal axis of inertia. I_{zz} is the moment of inertia of the block + bullet system after the collision and can be written as

$$I_{zz} = I_{z,B} + md_b^2 + Md_B^2, \tag{14.20}$$

where $I_{z,B}$ is the moment of inertia of the block about an axis parallel to z that goes through B, and d_B and d_b are the distances, after the collision, between the center of mass of the system and, respectively, point B and the bullet. Using the relation $Md_B = md_b$, one obtains,

$$I_{zz} = M\left(k_B^2 + \alpha d_f^2\right), \tag{14.21}$$

where $d_f \equiv d_b + d_B$ is the distance between the bullet and the center of mass of block B after the collision and $k_B \equiv \sqrt{I_{z,B}/M}$ is the radius of gyration of the block. Therefore we can conclude that all the details of the collision are contained in the (length) parameters d_f and k_B: they depend on how exactly the bullet deforms the block and affects its mass distribution and where exactly the bullet stops inside the block. These lengths cannot be predicted by the conservation laws we are using; an estimate of their values requires independent modeling of the interactions between the bullet and the block during collision (much like the "Hertz contact" that is used in elastic collisions—see e.g. [5]).

Using Eqs. (14.19) and (14.21), the fraction of the initial energy that is transformed into kinetic rotational energy is,

$$\frac{K_r}{E_i} = \frac{\alpha(1-\alpha)d_i^2 \sin^2\theta}{k_B^2 + \alpha d_f^2}. \tag{14.22}$$

We will now proceed to analyze this expression for two particular types of block: (i) a thin rod and (ii) a homogeneous rectangular parallelepiped, the situation closer to the experiments of [2–4].

14.3.1 Thin rod

Let us consider that the block is a thin rod of length ℓ, not necessarily homogeneous. The bullet will stick to the rod at the point of impact so that $d_i = d_f = d$. For simplicity, we will analyze only the case where $\vec{v}$ is perpendicular to the rod (i.e. $\theta = \pi/2$). The rotational kinetic energy becomes,

$$\frac{K_r}{E_i} = \frac{\alpha(1-\alpha)a^2}{1+\alpha a^2}, \tag{14.23}$$

where $a = d/k_B$ is the ratio of the distance of the bullet to the CM of the rod and the radius of gyration of the rod. A large value of a means either that the bullet hits the block far away from its CM, or that the mass of the rod is concentrated at its midpoint; $a = 0$ corresponds to $L_{CM} = 0$. For a homogeneous rod, a has a maximum possible value, $a_{max} = \sqrt{3}$, because $K_B = \ell/\sqrt{12}$ and the maximum value of d is $\ell/2$.

In Figs. 14.2 and 14.3 we analyze Eq. (14.23). For a fixed value of α, K_{rot}/E_i is an increasing function of a, with a maximum value of $1 - \alpha$ obtained in the limit $a >> 1$. This limit corresponds to having no energy dissipation: a fraction α of E_i is transformed into translational kinetic energy (see Eq.(14.17)) and a fraction $1 - \alpha$ into rotational kinetic energy. For low values of a (see the inset of Fig. 14.2), K_{rot}/E_i is larger for intermediate values of α. This is confirmed in Fig. 14.3, where K_{rot}/E_i is plotted as a function of α: for each value of a there is a value of α for which K_{rot}/E_i is a maximum. The value of this maximum increases when a increases. There are, therefore, two main ways of increasing the rotational kinetic energy and thus decreasing the dissipated energy: (i) decrease the radius of gyration of the system by concentrating the mass of the rod in its center and (ii) increasing m relatively to M for fixed values of a.

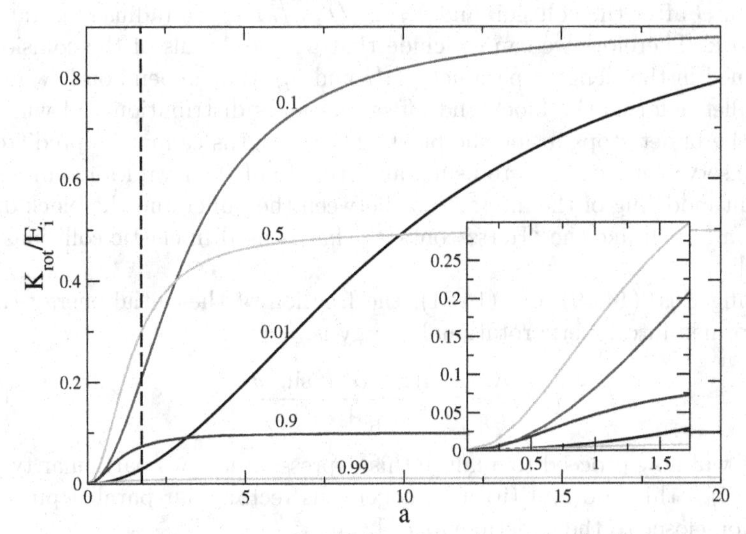

FIGURE 14.2
Fraction of initial mechanical energy transformed into rotational energy as a function of a, the ratio between d (the distance from the center of mass of the thin rod to the bullet) and k_B (the radius of gyration of the thin rod), for the indicated values of α. The dashed line represents $a_{max} = \sqrt{3}$, the maximum possible value of a for a homogeneous rod. The inset represents the results for $a \leq a_{max}$.

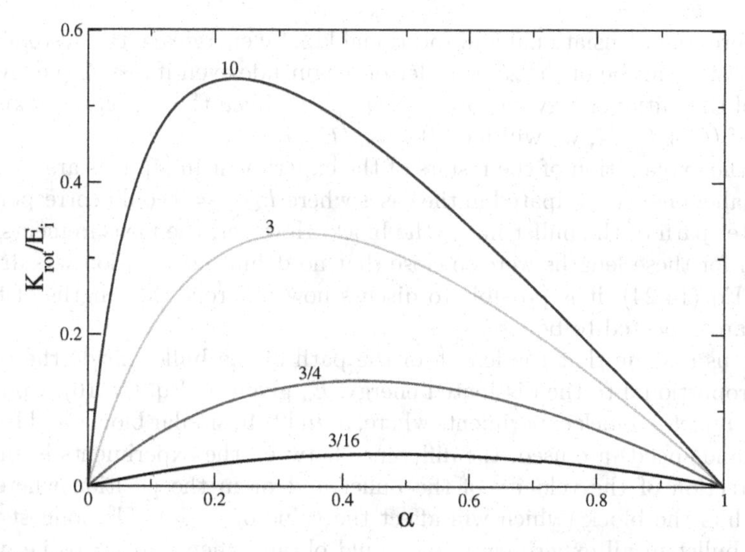

FIGURE 14.3
Fraction of initial mechanical energy transformed into rotational energy as a function of $\alpha \equiv m/(m+M)$, for the indicated values of a^2. $a^2 = 3$, $3/4$, and $3/16$ represent the cases where the bullet hits vertically a homogeneous rod of length ℓ at distances from its center of mass B equal to $\ell/2$, $\ell/4$, and $\ell/8$, respectively.

14.3.2 Rectangular parallelepiped

In this sections, we analyze a case that is as close as possible to the experiments shown in [2–4]. The block is a homogeneous rectangular parallelepiped whose edges have lengths (ℓ_x, ℓ_y, ℓ_z). The direction of the angular momentum is parallel to the edges of length ℓ_z, and the bullet hits the block in a face whose edges have lengths ℓ_x and ℓ_z. On the other hand, the mass of the bullet m is much smaller than the mass of block M, so in the analysis of this case we will only consider the limit $\alpha \ll 1$. Assuming that the radius of gyration of the block is unaffected by the collision, the rotational kinetic energy after the collision is then given by Eq. (14.22) with $k_B = \sqrt{(\ell_x^2 + \ell_y^2)/12}$, and its linear expansion around $\alpha = 0$ is,

$$\frac{K_r}{E_i} = 12\alpha \frac{d_i^2 \sin^2\theta}{\ell_x^2 + \ell_y^2}. \tag{14.24}$$

Notice that, to first order in α, K_r is independent of the final position of the bullet inside the block (i.e. of d_f). The ratio of the translational and rotational kinetic energies of the system after the collision is,

$$\frac{K_r}{E_{CM,f}} = 12 \frac{d_i^2 \sin^2\theta}{\ell_x^2 + \ell_y^2}. \tag{14.25}$$

Therefore, the translational and rotational kinetic energies of the system after the collision can be of the same order of magnitude even if $\alpha \ll 1$ (i.e. even if most of the initial energy has been dissipated). Given that d_i has a maximum value of $(\ell_x^2 + \ell_y^2)/4$, we will have $0 < K_r/E_{CM,f} < 3$.

In the explanation of the results of the experiment in [4], it is argued that the smaller energy dissipated in the cases where $L_{CM} \neq 0$ could correspond to a shorter path of the bullet inside the block. However, the measurements performed for these lengths were so close that no definite conclusion was drawn. Using Eq. (14.24), it is possible to discuss how different the lengths of these paths are expected to be.

Let us assume that the length of the path of the bullet inside the block, s, is proportional to the dissipated energy, E_d given by Eq. (14.16). Consider several bullet + block experiments where, as in [2–4], similar blocks and bullets (mass and speed) are used; the difference between the experiments is only in the direction of the velocity of the bullet and/or in the position where the bullet hits the block (which will affect the value of L_{CM}). The longest path of the bullet in all experiments, s_0, would obtain when $L_{CM} = 0$ (i.e. when $d_i \sin\theta = 0$), with $s_0 \propto E_{d,0} \equiv E_i(1-\alpha)$. In general, one can compare a length s of any experiment with s_0 using,

$$\Delta = \frac{s_0 - s}{s_0} = \frac{E_{d,0} - E_d}{E_{d,0}} = \frac{K_r}{E_i(1-\alpha)} \approx \frac{K_r}{E_i}. \tag{14.26}$$

In the experimental setup of [2–4], where α is fixed and the blocks are all equal, only d_i and θ can be changed in order to increase Δ. Δ has a maximum possible value since $d_i^2 \leq (\ell_x^2 + \ell_y^2)/4$. Therefore, the maximum expected deviation in the length of the different paths of the bullet, obtained by varying the position of the point of impact and/or the direction of the bullet's velocity, is $\Delta_{max} = 3\alpha$. For lengths s of the order of centimeters, a value of $\Delta \geq 0.1$ would give differences of the order of millimeters that could be measured experimentally, and this is obtained for $m/M >\approx 0.03$.

14.4 Conclusions

Inspired by some popular experiments presented on the physics YouTube channel *Veritasium*, we have revisited the problem of the collision between a bullet and a rigid body, in the case where there is no fixed point in the system (contrary to what is more often considered in basic textbooks). We have shown that the velocity of the CM of the system after the collision is independent of its angular momentum about the CM and that the energy dissipated in the collision decreases when the angular momentum about the CM increases. These results explain rigorously the two main questions raised by the experiments of the videos [2–4]: (i) why do the blocks reach the same height when fired with equal bullets having equal velocities, irrespectively of the position where the bullet hits the block; (ii) why does the dissipated energy decreases when the distance between the point where the bullet hits the block and the center of mass of the block is increased. Question (i) was answered in [4] using linear momentum conservation. Here we have clarified that even if the effect of gravity during the collision is not neglected, the result still holds. The answer to question (ii) was given by deriving the relation between the dissipated energy and the angular momentum, which shows that the dissipated energy decreases when angular momentum increases. Finally, we have analyzed collisions with two particular types of blocks. In the case of a thin rod, the analysis is simplified because one has to assume that the bullet sticks to the block (instead of traveling inside it). This example allows us to understand some instructive limiting cases (despite of the difficulty to design them experimentally): if the mass of the rod is concentrated at its midpoint, then the dissipated energy can be decreased to values close to 0; for any given mass distribution of the rod and distance of impact of the bullet, an intermediate value for the mass of the bullet can be chosen so that the fraction of initial energy transformed into rotational kinetic energy is maximum. The case of a homogeneous rectangular parallelepiped was studied in the limit where the mass of the bullet is much smaller than the mass of the block. It was shown that the ratio of the rotational and translational kinetic energies after the collision has an upper limit of 3 and that the relative difference in the length of the paths of bullets inside the block scales (if one assumes that the dissipated energy is proportional to this length) with the ratio of bullet and block masses. Therefore, the experimental test of this possible relation requires the use of heavier bullets.

Acknowledgments

I thank my students for presenting me the questions posed in [2]. It was their interest that originated this work.

References

[1] http://www.youtube.com/user/1veritasium

[2] http://www.youtube.com/watch?v=vWVZ6APXM4w

[3] https://www.youtube.com/watch?v=N8HrMZB6_dU

[4] http://www.youtube.com/watch?v=BLYoyLcdGPc

[5] P. Patrício, "The Hertz contact in chain elastic collisions," Am. J. Phys. **72**, 1488–1491 (2004).

[6] P. M. Fishbane, S. Gasiorowicz, and S. T. Thornton, *Physics for Scientists and Engineers*, 3rd ed. (Pearson Prentice-Hall, Upper Saddle River, NJ, 2005).

15

The Continuity Equation in Ampère's Law

J. P. Silva and A. J. Silvestre

CONTENTS

We use a simple problem to illustrate some subtleties of the relationship between the Biot-Savart law and Ampère's law.

15.1 Introduction

When learning electromagnetism students are often told that very similar techniques are used to compute both the electric field $\vec{E}$ and the magnetic flux density $\vec{B}$. For example, the $\vec{E}$ field of a continuous charge distribution can be computed using the definition of an electrostatic field that follows from Coulomb's law, i.e., the electrostatic force divided by the "test charge". By analogy, one can define the $\vec{B}$ field of distribution of steady currents using the Biot-Savart law. Likewise, for very symmetric charge distributions, one may use Gauss' law to find the $\vec{E}$ field via the electrostatic potential and Ampère's law to find $\vec{B}$. Finally, the electrostatic field can be computed by first finding the electrostatic potential, and the magnetic flux density by first finding the magnetic vector potential. The analogy between methods for finding fields $\vec{E}$ and $\vec{B}$ thus seems perfect.

Students are also often imparted the notion that any of the calculational techniques outlined above can, in principle, be used in any situation, although some techniques may be easier to apply than others depending on the specific problem at hand. This leads to a hierarchy of the above techniques. For

DOI: 10.1201/9781003187103-15

example, when computing $\vec{E}$ of a given charge distribution, the principle of least effort dictates that we start by inspecting the problem for symmetries that may allow us to use Gauss' law. If not, the next thing to try is to compute the electrostatic potential and get the field as (minus) its gradient. Indeed, because $\vec{E}$ is a vector, computing it directly is not always worth the effort, even if the symmetry is low (except, of course, in an assessment, as the instructor is then free to request whatever pleases them...) [1].

Students are thus led to believe there exists a perfect analogy between techniques for finding $\vec{E}$ and $\vec{B}$ [2]: in either case there are three methods, all of which are generally applicable (and which should be attempted in increasing order of foreseeable operational difficulty).

In this article, we discuss the computation of the magnetic flux density inside a square loop of wire carrying a steady current. We show that this example exposes some subtleties of the two notions described above. This is because our problem can be broken down into simpler problems, each of which, in and of itself, is physically meaningless. In particular, our example makes it apparent that:

1. Field $\vec{B}$ cannot be found by applying Ampère's law to individual pieces of circuit, unlike what one might think could be proved. One must resort to the Biot-Savart law instead.
2. Deriving Gauss' law from Coulomb's law is not an exact analog of deriving Ampère's law from the Biot-Savart law.

15.2 The problem and its electrostatic analog

Let us consider a square loop of wire, of side length L and carrying a steady current I. Our aim is to find field $\vec{B}$ at an arbitrary position inside the loop (see Fig. 15.2a). A roughly similar electrostatics problem is: find the field $\vec{E}$ of a square loop of side length L if two adjacent sides of the loop have charge $+2\lambda L$, and the other two sides have charge $-2\lambda L$, where the charges are uniformly distributed with linear density λ (see Fig. 15.2b). The similarity between the two problems is, of course, not perfect. For example, it follows straightforwardly from Coulomb's law that field $\vec{E}$ lies in the plane of the loop in Fig. 15.2b, whereas from the Biot-Savart law, it follows that field $\vec{B}$

[1] There is a difference between $\vec{E}$ and $\vec{B}$ here: because the magnetic potential is a vector, it is not necessarily cheaper to compute first $\vec{A}$ and then $\vec{B} = \text{curl}\,\vec{A}$, than to find $\vec{B}$ directly from the Biot-Savart law. We shall not consider applications of the vector potential in magnetostatics, for which the analogy is better.

[2] Some textbooks are even laid out so as to highlight this analogy. See, e.g., A. Shadowitz, *The Electromagnetic Field*, Dover Publications, Inc., New York, 1988.

is perpendicular to the plane of the loop in Fig. 15.2a. But there is another, much more subtle difference.

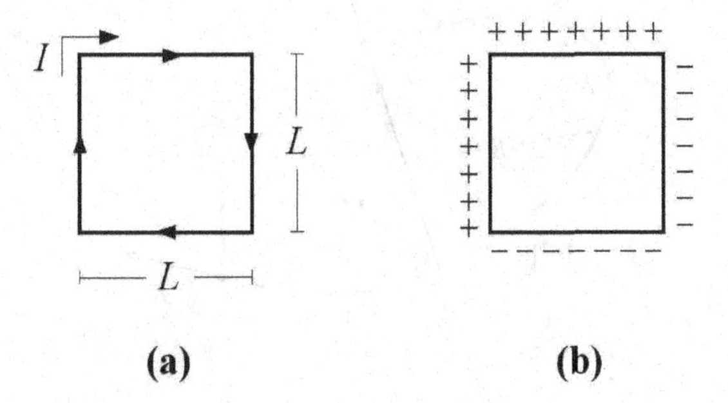

FIGURE 15.1
(a) Square loop of side length L carrying a steady current I. (b) Square loop of side length L with electric charge λL on two of its sides and charge $-\lambda L$ on the two remaining sides.

Let us proceed with Fig. 15.2b. Obviously, we cannot use Gauss' law, as the charge distribution is not symmetric enough. So, with a view to invoking the principle of superposition, we start by focussing on the field of linear distribution of charge of length L, with uniform linear charge density λ. We next look for a surface (call it "gaussian surface") such that the dot product of $\vec{E}$ with the unit normal to that surface, $\vec{n}$, is the same at every point on the surface. It is not at all easy to guess what shape such a surface should be, so we resort to Coulomb's law or the electrostatic potential as a route to $\vec{E}$. All that remains to do is write down the results for all four sides of the loop—which differ only in their distance to some generic point P at which we want to find $\vec{E}$—and add them together.

Let us now tackle the problem of finding field $\vec{B}$ of the steady current carried by the loop in Fig. 15.2a. We likewise acknowledge that the symmetry is too low to use Ampère's law. So we start by breaking up the loop into four straight wires of length L, each carrying current I. Next, we enquire whether we might use Ampère's law to find the $\vec{B}$ field of each individual wire (which, for simplicity, we assume to lie along the z-axis). To this end, we try to guess what shape a closed curve (call it "amperian path") should be such that the dot product of $\vec{B}$ with the unit vector tangent to the curve, $d\vec{l}/||d\vec{l}||$, is the same at every point on the curve. Consider first a point on the perpendicular bisector plane of the wire. Let the point be a distance R from the wire (see Fig. 15.2). From the cross product in the Biot-Savart law, we straightforwardly conclude that field $\vec{B}$ at this point is perpendicular to both the wire and the line connecting the point to the wire, as is the infinitesimal arc length along the circle of radius R, $d\vec{l}$. By symmetry, however, field $\vec{B}$ will have the same magnitude at all points in this plane located the same distance R from the

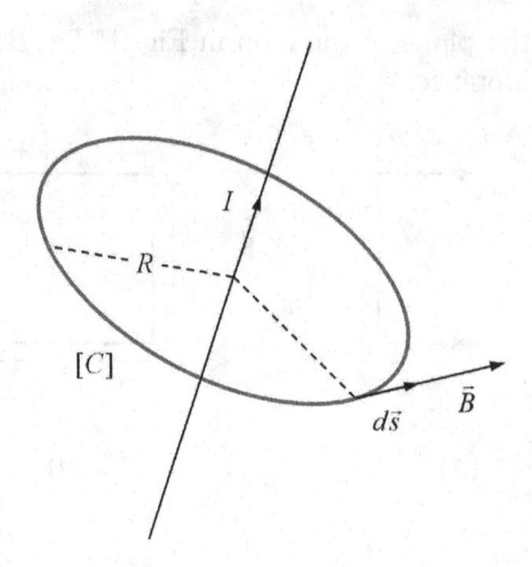

FIGURE 15.2
Schematic for calculating the magnetic flux density of a straight wire carrying a steady current, using Ampère's law.

wire. It is then apparent that $\vec{B} \cdot d\vec{l}/||d\vec{l}||$ is the same at every point of a circle of radius R lying in this plane and centered on the wire. We can thus apply Ampère's law to this circle, with the result

$$\int \vec{B} \cdot d\vec{l} = \mu_0 I \Rightarrow B \int dl = \mu_0 I, \tag{15.1}$$

whence $B = \mu_0 I/2\pi R$.

Having done that, some questions remain. On the one hand, the result we obtained does not depend on the length of the wire, which is rather odd. On the other hand, if we repeat the same reasoning for any other point a distance R from the wire but with z between $-L/2$ and $+L/2$, we will always get $B = \mu_0 I/2\pi R$. And yet if we repeat the calculation for a point with $z < -L/2$ or $z > L/2$, we will get 0, because in this case the current-carrying wire does not pass through the plane circular area bounded by the circle of radius R. Something does not seem right...[3] To play safe, let us re-calculate $\vec{B}$ using the Biot-Savart law (see Fig. 15.2).

[3]Things may be even more serious. Indeed, the term on the right-hand side of Ampère's law is proportional to the current that passes through any surface bounded by the amperian path (see, e.g., M. H. Nayfeh and M. K. Brussel, *Electricity and Magnetism* (Wiley International, 1985), p. 252). Now it is easily seen that we can deform this surface in such a way that the current either passes or does not pass through it, as we wish. Clearly, this is an ambiguous result.

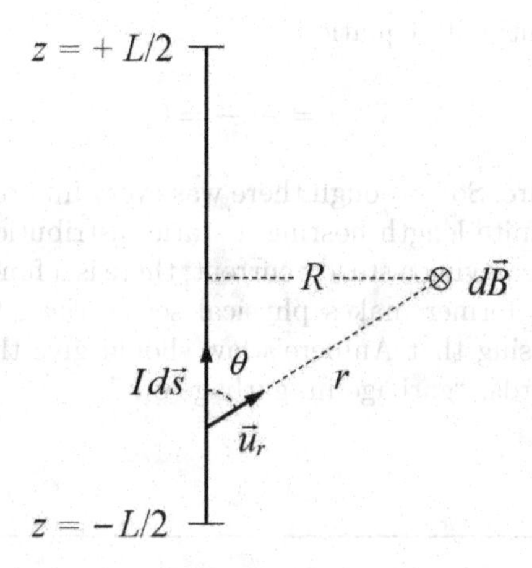

FIGURE 15.3
Schematic for calculating the magnetic flux density of a straight wire carrying a steady current, using the Biot-Savart law.

Again, we start by considering a point on the perpendicular bisector of the wire and a distance D from the wire. From the Biot-Savart law,

$$||\vec{B}|| = \frac{\mu_0}{4\pi} I \frac{||d\vec{s} \times \vec{u}_r||}{r^2}, \tag{15.2}$$

we find

$$\begin{aligned} ||\vec{B}|| &= \int d||\vec{B}|| = \frac{\mu_0 I}{4\pi} \int \frac{||d\vec{s} \times \vec{u}_r||}{r^2} = \frac{\mu_0 I}{4\pi} \int \frac{ds \sin\theta}{r^2} \\ &= \frac{\mu_0 I}{4\pi} \int_{-L/2}^{L/2} \frac{R dz}{(R^2 + z^2)^{3/2}} = \frac{\mu_0 I}{2\pi R} \frac{L}{\sqrt{4R^2 + L^2}}, \end{aligned} \tag{15.3}$$

and we have blotted our copybook! This result is not the same as we got from Ampère's law [4].

We are in a quandary. On the one hand, the Biot-Savart law holds for any steady current, and we have done our sums right. On the other hand, we also know that Ampère's law holds for any steady current, and we have chosen a convenient amperian path. What have we done wrong? The problem is that a wire of finite length carrying a steady current is a meaningless concept. There just is no such thing as a wire of finite length with current going in at $z = -L/2$ and coming out at $z = L/2$. Where would the charges come from, and where would they go? In actual fact, the finite-length current-carrying

[4] Another way to see this is to note that integrating the $\vec{B}$ field given by the Biot-Savart law along a circle perpendicular to the wire and centered on it does not give $\mu_0 I$.

wire violates the continuity equation

$$\nabla \cdot \vec{J} = -\frac{\partial \rho}{\partial t} = 0, \tag{15.4}$$

at the ends of the wire. So, although there was every hint of a strong similarity between a wire of finite length hosting a static distribution of charges and a wire of finite length carrying a steady current, there is a fundamental difference between them. The former makes physical sense, the latter does not. It is therefore not surprising that Ampère's law should give the wrong result. To quote some wise words, "garbage in, garbage out".

15.3 The difference between the Biot-Savart law and Ampère's law

There is a subtle difference between the Biot-Savart law and that of Ampère. Indeed, deriving Ampère's law from the Biot-Savart law requires one additional ingredient: the continuity equation [1].

Ampère's law incorporates the continuity equation and thus "knows" that the magnetic flux density of a finite wire is a meaningless concept. So, if we use it to find this $\vec{B}$, we get the wrong result. On the other hand, as we already mentioned, applying Ampère's law to the (meaningful) original square loop does not allow us to find $\vec{B}$, because we cannot guess what shape an amperian path needs to be along which $\vec{B} \cdot d\vec{l}/||d\vec{l}||$ is constant. So Ampère's law is not useful for solving this problem.

And what is the usefulness of the Biot-Savart law? If we combine the superposition principle with the (physically meaningless) $\vec{B}$ for the finite wire given by the Biot-Savart law, do we get the correct $\vec{B}$ for the square loop? Or will only garbage come out if garbage goes in? It is clearly possible to integrate $d\vec{B}$ along the square loop and this will necessarily give the correct result—the law holds, and the question we are asking is physically meaningful. But this integral can be split into four integrals, one along each side of the loop. We conclude that combining the principle of superposition with the Biot-Savart law applied to a finite wire does indeed yield the correct result, even though there is a physically meaningless intermediate step involved [2].

This difference in the applicability of the Biot-Savart and Ampère laws becomes just as subtly apparent when contrasting them with the Coulomb and Gauss laws. Indeed, Gauss' law is a direct consequence of Coulomb's law without any additional physical assumptions [3], so Gauss' law will always confirm the result of any calculation performed using Coulomb's law. In contrast, as we saw above, to derive Ampère's law from the Biot-Savart law, we must use the continuity equation. This is the reason why we can apply the Biot-Savart law to a wire of finite length (and then combine results to arrive at the correct result for $\vec{B}$ of a system that is physically meaningful as a whole), but not Ampère's law.

15.4 Conclusions

In this article, we used a very simple example to illustrate two subtle issues. On the one hand, that the derivation of Ampère's law from the Biot-Savart law requires the continuity equation as an additional physical assumption, in contrast to Gauss' law which is a direct consequence of Coulomb's law. On the other hand, because Ampère's law incorporates the continuity equation and the Biot-Savart law does not, Ampère's law when applied to a physically meaningless problem will flag it as such (by providing an incorrect result for $\vec{B}$), whereas the Biot-Savart law will yield a partial result that can be added to other partial results to recover the correct $\vec{B}$ for a system that is physically meaningful as a whole.

It cannot be overemphasized that what we have done does not in any way restrict the validity of Ampère's law. It is actually the Biot-Savart law (and not Ampère's) that is of restricted validity. Indeed, Ampère's law when supplemented with the displacement current term is one of Maxwell's four equations. Our original problem, to find the magnetic flux density of a square loop of wire carrying a steady current, is physically meaningful and lies well within the range of applicability of either laws. Both the Biot-Savart law or Ampère's law hold. By breaking up the loop into finite-length pieces, we are focussing on a *physically meaningless* problem. Both laws are still valid, the problem is not. Ampère's law "knows" this and applying it "blindly" gives us the wrong answer. In contrast, the Biot-Savart law does not incorporate the continuity equation, so it "does not realise" that the problem is meaningless. We can apply it "blindly" to reconstruct the original problem and find the correct result for the magnetic flux density.

Acknowledgments

We are grateful to A. Barroso, M. R. Gomes, and R. Santos for a careful reading of, as well as many comments and suggestions on, the manuscript

References

[1] J. D. Jackson, *Classical Electrodynamics*, 2nd ed. (John Wiley and Sons, 1975), pp. 173–175.

[2] In this connection the footnote on page 170 of reference [1] is very interesting to read: it discusses how the integral form of the Biot-Savart law for circuits is related to the magnetic field of a moving charge.

[3] See reference [1], pp. 30–32.

Part III

Misconceptions

16

On the Relation between Angular Momentum and Angular Velocity

J. P. Silva and J. M. Tavares

CONTENTS

Students of mechanics usually have difficulties when they learn about the rotation of a rigid body. These difficulties are rooted in the relation between angular momentum and angular velocity because these vectors are not parallel, and we need in general to utilize a rotating frame of reference or a time dependent inertia tensor. We discuss a series of problems that introduce both difficulties.

16.1 Introduction

Students of introductory or intermediate level mechanics courses have a notorious difficulty in tackling problems involving rotation. Part of their difficulty is conceptual, and it comes from the fact that the laws governing translation

Reproduced from J. P. Silva and J. M. Tavares, "On the relation between angular momentum and angular velocity", American Journal of Physics **75**, 53–55 (2007), https://doi.org/10.1119/1.2388970, with the permission of the American Association of Physics Teachers.

and rotation are deceptively similar. We write

$$\boldsymbol{F} = \frac{d\boldsymbol{P}}{dt}, \quad (16.1)$$

$$\boldsymbol{\tau_O} = \frac{d\boldsymbol{L_O}}{dt}, \quad (16.2)$$

where $\boldsymbol{F}$ is the total external force, $\boldsymbol{P}$ is the linear momentum of the system, $\boldsymbol{\tau_O}$ is the total external torque about the point O, and $\boldsymbol{L_O}$ is the angular momentum of the system about O. We have assumed that the origin of the coordinate system is not accelerating and that its axes do not rotate[1].

There are two important differences between the two equations [1]. We obtain the equations of motion from Eq. (16.1) by substituting $\boldsymbol{P} = m\boldsymbol{V}$, where m is the total mass of the system and $\boldsymbol{V}$ is the velocity of the center of mass. The analysis is simplified by the fact that $\boldsymbol{P}$ is parallel to $\boldsymbol{V}$. In contrast, $\boldsymbol{L_O}$ is not parallel to $\boldsymbol{\omega}$ in general, and their relation is given by

$$\begin{pmatrix} L_{Ox} \\ L_{Oy} \\ L_{Oz} \end{pmatrix} = \begin{pmatrix} I_{Oxx} & I_{Oxy} & I_{Oxz} \\ I_{Oyx} & I_{Oyy} & I_{Oyz} \\ I_{Ozx} & I_{Ozy} & I_{Ozz} \end{pmatrix} \begin{pmatrix} \omega_x \\ \omega_y \\ \omega_z \end{pmatrix}. \quad (16.3)$$

We also can always choose any inertial frame of reference to bring the translational dynamics into the form of Eq. (16.1). But, the choice of an inertial frame of reference is usually not advisable when studying rotations, because it would force the coefficients of the matrix I_O to vary with time. Conversely, if a time-independent I_O is defined, it is not always possible to describe the rotation by the simple form of Eq. (16.2).

We have found that our students learn better when these difficulties are first introduced with the help of the sequence of problems we present in the following. The problems are simple and involve one or two-point particles in circular motion.

16.2 Angular momentum of a particle describing circular motion

Consider a point particle of mass m in free space, which moves in a circle of radius R around the point $a\hat{\boldsymbol{x}} + b\hat{\boldsymbol{y}} + c\hat{\boldsymbol{z}}$, with a constant angular velocity

[1] We can always make this choice when studying Eq. (16.1). We may depart from such a choice, for instance, when we wish to study what a particular motion will look like as viewed from an accelerated frame (as in a passenger traveling on a bus coming to rest near a bus stop; or a very long Foucault pendulum which swings under the influence of the rotating earth). In these cases, Eq. (16.1) must be modified to include the inertial forces. But, we may always choose to study the translational dynamics of these examples in an inertial frame reference, where the inertial forces are absent.

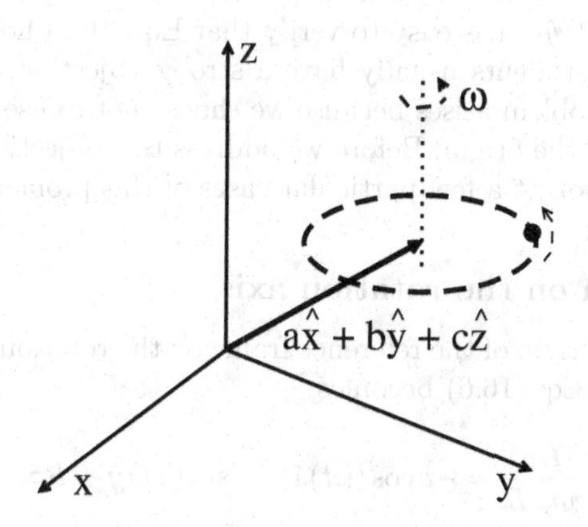

FIGURE 16.1
Point particle of mass m describing a circular motion of radius R centered at $a\hat{\boldsymbol{x}} + b\hat{\boldsymbol{y}} + c\hat{\boldsymbol{z}}$ and revolving around the z-axis with constant angular velocity ω.

$\boldsymbol{\omega} = \omega\hat{\boldsymbol{z}}$, as shown in Fig. 16.1. With a suitable choice for the origin of time t, the motion is described by the position vector

$$\boldsymbol{r}_O = [a + R\cos(\omega t)]\,\hat{\boldsymbol{x}} + [b + R\sin(\omega t)]\,\hat{\boldsymbol{y}} + c\hat{\boldsymbol{z}}. \tag{16.4}$$

For the time derivative we find

$$\frac{\boldsymbol{v}_O}{\omega R} = -\sin(\omega t)\hat{\boldsymbol{x}} + \cos(\omega t)\hat{\boldsymbol{y}}, \tag{16.5}$$

and the angular momentum about point O, $\boldsymbol{L}_O = \boldsymbol{r}_O \times (m\boldsymbol{v}_O)$, becomes

$$\begin{aligned} \frac{\boldsymbol{L}_O}{m\omega R} &= \frac{\boldsymbol{r}_O \times \boldsymbol{v}_O}{\omega R} \\ &= -c\cos(\omega t)\hat{\boldsymbol{x}} - c\sin(\omega t)\hat{\boldsymbol{y}} + [a\cos(\omega t) + b\sin(\omega t) + R]\,\hat{\boldsymbol{z}} \end{aligned} \tag{16.6}$$

Note that for the generic case described thus far, neither the magnitude of $\boldsymbol{L}_O$ nor its z component are constant.

We recall that $\omega_x = 0 = \omega_y$, and $\omega_z \equiv \omega$, and write the moment of inertia in the form of Eq. (16.3) by setting

$$\begin{aligned} I_{O\,xz} &= -mRc\cos(\omega t), \\ I_{O\,yz} &= -mRc\sin(\omega t), \\ I_{O\,zz} &= mR\,[a\cos(\omega t) + b\sin(\omega t) + R]\,. \end{aligned} \tag{16.7}$$

These equations highlight the two problems mentioned in Sec. 16.1. Namely, that $\boldsymbol{L}_O$ is not parallel to $\boldsymbol{\omega}$, and that if we choose a generic inertial frame, I_O becomes time-dependent. If we use $\mathbf{F} = -m\omega^2 R$ $[\cos(\omega t)\hat{\boldsymbol{x}} + \sin(\omega t)\hat{\boldsymbol{y}}]$, it is easy to verify that Eq. (16.2) holds in this case.

At this stage, students usually have a strong objection. They state (correctly) that the problem arises because we choose not to use the center of the circular motion as the origin. Before we address this objection in Sec. 16.3, it is interesting to look at a few particular cases of this problem.

16.2.1 Origin on the rotation axis

The choice of the origin of the reference frame on the rotation axis corresponds to $a = b = 0$, and Eq. (16.6) becomes

$$\frac{\boldsymbol{L}_O}{m\omega R} = -c\cos(\omega t)\hat{\boldsymbol{x}} - c\sin(\omega t)\hat{\boldsymbol{y}} + R\hat{\boldsymbol{z}}. \tag{16.8}$$

In this case the z component of the angular momentum, $L_{Oz} = mR^2\omega$ and its magnitude $L_O = L_{Oz}\sqrt{c^2 + R^2}/R$ are constant. The vector $\boldsymbol{L}_O$ precesses around $\boldsymbol{\omega}$, with its tip describing circular motion about the z-axis. Here the magnitude of $\boldsymbol{L}_O$ is constant, but its direction is not. This situation occurs in a variety of physical situations.

16.2.2 Origin on the plane of motion

The choice of the origin of the reference frame on the plane of motion corresponds to $c = 0$ and, without loss of generality, we can set b=0. In this case, Eq. (16.6) becomes

$$\boldsymbol{L}_O = mR\left[a\cos(\omega t) + R\right]\omega\hat{\boldsymbol{z}}, \tag{16.9}$$

and $\boldsymbol{L}_O$ becomes parallel to $\boldsymbol{\omega} = \omega\hat{\boldsymbol{z}}$. But the magnitude of $\boldsymbol{L}_O$ varies with time between a minimum of $mR\omega(R-a)$ and a maximum of $mR\omega(R+a)$. Here the direction of $\boldsymbol{L}_O$ is constant, but its magnitude is not.

16.2.3 Origin on the center of circular motion

In this case $a = b = c = 0$, and we recover the well known result $\boldsymbol{L}_O = mR^2\boldsymbol{\omega}$. These examples stress that if we state that a central force leads to conservation of angular momentum, we must specify that the angular momentum is calculated about the center of motion.

The choice $a = b = c = 0$ supports the students' objection that the complications in Eq. (16.7) have arisen from a bad choice for O. The next example shows that a simple alteration makes it impossible to find an ideal choice for O.

16.3 Angular momentum of two particles describing circular motion

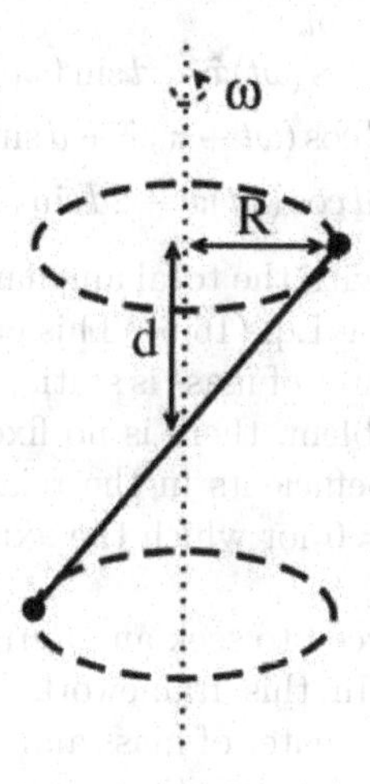

FIGURE 16.2
Point particles of mass m describing a circular motion of radius R, centered at a common axis with constant angular velocity ω. The particles are always on opposite sides of the axis.

Consider the two particles shown in Fig. 16.2, which rotate in free space around a common axis z, with the same angular velocity ω, in two parallel planes at a vertical distance $2d$ from each other. The particles rotate such that they always lie on opposite sides of the axis (*i.e.*, the two particles are out of phase by π). The system as a whole may be viewed as a dumbbell rotating around an axis which makes an angle $\arctan(R/d)$ with the axis of the dumbbell.

The total angular momentum of the system about a point O is given by the sum of the two angular momenta, each calculated in Sec. 16.2 for a generic point O. For the lower particle (particle 2) we could choose O to coincide with the center of its circular orbit, thereby simplifying its angular momentum to be $\boldsymbol{L}_O = mR^2\boldsymbol{\omega}$. The result for the angular momentum of the upper particle (particle 1) is given by Eq. (16.8) with $c = 2d$. Alternatively, we could choose O to coincide with the center of particle 1's circular orbit. This choice would simplify its angular momentum but then, the angular momentum of particle 2 would be given by Eq. (16.8) with $c = -2d$ and $\omega t \to \omega t + \pi$. In either description the total angular momentum is not parallel to $\boldsymbol{\omega}$ and the inertia tensor depends on time.

We might also choose O along the rotation axis, in the middle between the two planes of motion. This choice coincides with the center of mass of the system, as shown in Fig. 16.2. Then, the two angular momenta are given by Eq. (16.8) with $c = d$ and $c = -d$ $(\omega t \to \omega t + \pi)$, leading to

$$\begin{aligned}\frac{\boldsymbol{L}_{\text{cm}}}{m\omega R} &= \frac{\boldsymbol{L}_{1\ \text{cm}} + \boldsymbol{L}_{2\ \text{cm}}}{m\omega R} \\ &= -d\cos(\omega t)\hat{\boldsymbol{x}} - d\sin(\omega t)\hat{\boldsymbol{y}} + R\hat{\boldsymbol{z}} \\ &\quad + d\cos(\omega t + \pi)\hat{\boldsymbol{x}} + d\sin(\omega t + \pi)\hat{\boldsymbol{y}} + R\hat{\boldsymbol{z}} \\ &= -2d\cos(\omega t)\hat{\boldsymbol{x}} - 2d\sin(\omega t)\hat{\boldsymbol{y}} + 2R\hat{\boldsymbol{z}}. \end{aligned} \tag{16.10}$$

Equation (16.10) coincides with the total angular momentum about any point in space, as can be seen using Eq. (16.6). This curious result is a consequence of the fact that when the center of mass is static, $\boldsymbol{L}_O = \boldsymbol{L}_{\text{cm}}$ for any fixed point O. As a result, for this problem, there is no fixed point in space that makes $\boldsymbol{L}$ parallel to $\boldsymbol{\omega}$, nor the coefficients in the relation between $\boldsymbol{L}$ and $\boldsymbol{\omega}$ time-independent (except for $d = 0$ for which the axis of rotation is perpendicular to the axis of the dumbbell).

Consequently, we are forced to seek an alternative framework to deal with rotations of a rigid body. In this framework, we first choose a coordinate system with its origin at the center of mass and with its axis rigidly connected to the body. With this choice, the inertia tensor I_O becomes time-independent but the axis rotate with the body. As a result, Eq. (16.2) no longer holds, and we must use instead

$$\boldsymbol{\tau}_{\text{cm}} = \left(\frac{d\boldsymbol{L}_{\text{cm}}}{dt}\right)_{\text{rotating}} + \boldsymbol{\omega} \times \boldsymbol{L}_{\text{cm}}, \tag{16.11}$$

where the time derivative refers to the rate of change of $\boldsymbol{L}_{\text{cm}}$ with respect to the frame rotating rigidly with the body.

At this point there are several interesting choices. We can stick to the inertial reference frame in Fig. 16.2, where the moments of inertia are time dependent, and use Eqs. (16.2) and (16.10) in order to calculate $\boldsymbol{\tau}$. This choice gives $2\boldsymbol{r}_1 \times \boldsymbol{F}_1$, with $\boldsymbol{r}_1 = R\cos(\omega t)\hat{\boldsymbol{x}} + R\sin(\omega t)\hat{\boldsymbol{y}} + d\hat{\boldsymbol{z}}$, and $\boldsymbol{F}_1 = -m\omega^2 [R\cos(\omega t)\hat{\boldsymbol{x}} + R\sin(\omega t)\hat{\boldsymbol{y}}]$. We find that

$$\boldsymbol{\tau} = 2mRd\omega^2 [\sin(\omega t)\hat{\boldsymbol{x}} - \cos(\omega t)\hat{\boldsymbol{y}}], \tag{16.12}$$

showing that torque is needed to maintain the rotational motion with constant angular velocity. This result is at odds with the dynamics of translation, where constant momentum requires a vanishing external force. Alternatively, we can choose a frame that rotates with the dumbbell. In this case, the inertia tensor is time-independent, and we must use Eq. (16.11).

We have found that by using this series of problems, students obtain a physical intuition that guides them through the subsequent derivations involved in the correct definition of the inertia tensor I_O, its diagonalization by choosing the principal axes, and the derivation of Eq. (16.11). Moreover, their increased physical intuition serves them well in a variety of specific problems.

Acknowledgments

The authors are very grateful to C. R. Leal for useful discussions and for carefully reading and commenting on this manuscript. They also thank their students for guiding them toward good teaching choices.

Reference

[1] See, for example, K. R. Symon, *Mechanics*, 3rd ed. (Addison-Wesley, Reading, MA, 1971).

17

A Very Abnormal Normal Force

J. P. Silva and A. J. Silvestre

CONTENTS

Does the line of action of the normal force acting on a body lying on a surface always pass through the center of mass of the body? We argue that it need not be so, and propose a method for teaching this that is both paedagogical and entertaining.

17.1 Introduction

The purpose of this article is to present a method for teaching students that the line of action of the normal force on a body lying on a surface need not pass through the center of mass (CM) of the body. The method is both paedagogical and entertaining. This issue arises, e.g., when studying a body on an inclined plane, and it should be discussed only after the dynamics of rotating bodies have been taught.

It is, of course, always possible to present the correct solution from the start. However, our experience is that many college students (most of whom have been exposed to the correct solution) will still draw the normal force

Reproduced from J. P. Silva and A. J. Silvestre, "Uma normal muito anormal," Gazeta de Física **23**, 13–16 (2000), with the permission of the Portuguese Physical Society (SPF).

DOI: 10.1201/9781003187103-17

on a body as acting on the body's CM. Surprisingly, a large fraction of these very same students will draw the friction force correctly—acting parallel to the surface and along the line of contact between body and surface (see Fig. 17.1).

This has led us to develop a game. In this game, students propose a number of versions of the free-body diagram for a body in static equilibrium on an inclined plane with friction. They then work through these diagrams with their instructor and find a succession of contradictions (see Fig. 17.1).

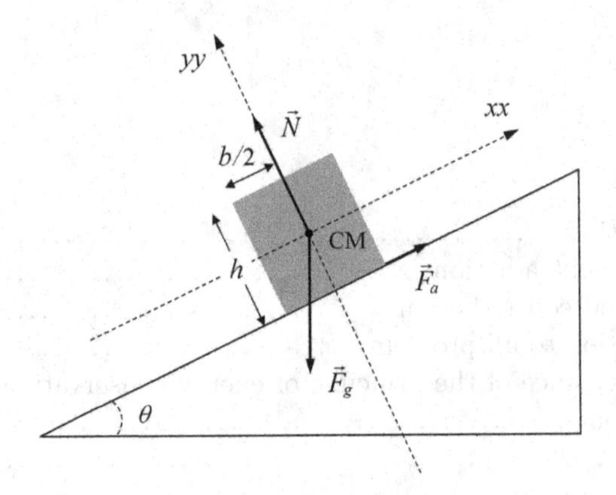

FIGURE 17.1
Free-body diagram for a body in static equilibrium on an inclined plane as often drawn by students. Can you tell what is wrong?

The advantages of this method, which we have used several times with some success, are:

1. It is a "fun" assignment, and therefore more easily remembered.
2. The reason why the free-body diagram of Fig. 17.1 is wrong becomes immediately apparent.
3. It illustrates the power of proof by contradiction.

17.2 The first contradiction

Let us consider a homogeneous body (e.g., an item of furniture) of height h and width b, placed on an inclined plane. Because of friction, the body does not slide down the plane [1]. Our purpose is to help students realize that the free-body diagram of Fig. 17.1 is wrong. We assume that students already know that:

(a) A force acting on a rigid body can be replaced by another force acting on some point along the original force's line of action. The two forces must have the same magnitude to have the same translational effect, and the same line of action to have the same rotational effect. Some textbooks express this as "a force can be slid along its line of action". In particular, the normal force could have been slid along its line of action so that it acts on the contact line between body and surface.

(b) The weight of the body must act on its CM. Indeed the net effect of gravity pulling downwards on all constituent particles of the body is a downward-pointing force acting on the body's CM.

(c) The friction force acts on and is parallel to the contact line between body and surface.

From Fig. 17.1, the following equation can be derived for the rotation of the body:

$$F_a \frac{h}{2} = I_{CM}\alpha, \tag{17.1}$$

where F_a is the magnitude of the friction force, I_{CM} is the moment of inertia of the body relative to its CM, and α is the angular acceleration. From the above equation, we find that

$$\alpha = \frac{F_a h}{2I_{CM}} \neq 0, \tag{17.2}$$

because F_a and h are both non-zero. It follows that the body rotates! In other words:

First contradiction: all furniture will topple!

For example, a wooden plank 100 m long and just 1 cm thick lying flat on the plane would tumble down.

Why is this wrong? We know that the weight and friction force are correctly accounted for. Nor do not seem to have missed any forces. We have no choice but to conclude that the only possible source of error is the assumption that the line of action of the normal force passes through the body's CM.

[1]This may arise in the context of many different problems in mechanics. For instance, what is the minimum static friction coefficient that will stop an item of furniture sliding down the plane? Or, how banked must a curved road be that a car traveling at a certain speed will round it without skidding down or turning over?

17.3 The second contradiction

In light of the above result, we propose to translate the line of action of the normal force a distance x to the left, so that the body will not necessarily rotate (see Fig. 17.3).

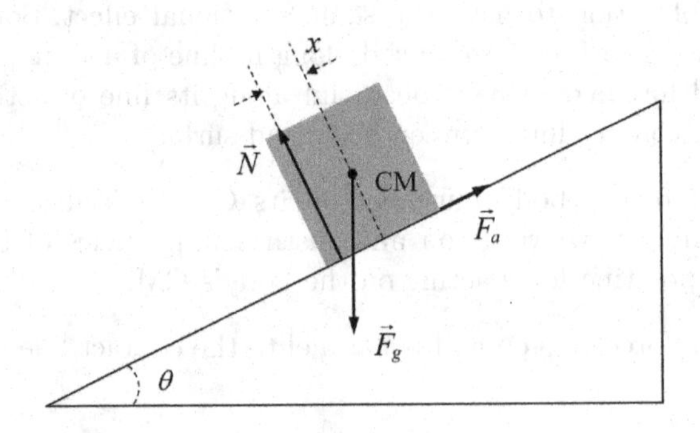

FIGURE 17.2
The line of action of the normal force has been translated a distance x to the left relative to the body's CM.

The equation for the rotation of the body is now

$$F_a \frac{h}{2} - Nx = I_{CM}\alpha, \tag{17.3}$$

which is solved by $\alpha = 0$ provided x satisfies

$$x = x_0 = \frac{F_a h}{2N} = \frac{hF_g \sin\theta}{2F_g \cos\theta} = \frac{h}{2}\tan\theta, \tag{17.4}$$

where in the third equality we used the fact that for a body in static equilibrium on a plane of inclination θ, the magnitudes of the normal force and the friction force are related to the body weight F_g via $N = F_g \cos\theta$ and $F_a = F_g \sin\theta$, respectively.

Now Eq. (17.4) can be solved. In fact, for given h and θ we can find x_0 for which the body will be in static equilibrium. But now we have another problem, as this equation always has a solution, i.e., we can find x_0 from Eq. (17.4) for any choice of h and θ. It thus appears that

Second contradiction: no furniture will ever topple!

For example, if we take the same wooden plank as before and stand it on the plane instead ($h = 100$ m, $b = 1$ cm), the above result leads us to conclude that it will not fall, however inclined the plane might be [2].

[2] At this point we should tell students that we have only proved that there is one

17.4 The solution to all problems

Contradiction 2 is easily solved if one appreciates what is the physical meaning of x_0. An elementary construction shows that x_0 is the separation between two lines, both perpendicular to the plane: one passing through the body's CM, and the other through the point where the line of action of the body's weight (a vertical line) intersects the contact line between the body and the inclined plane (see Fig. 17.4) [3].

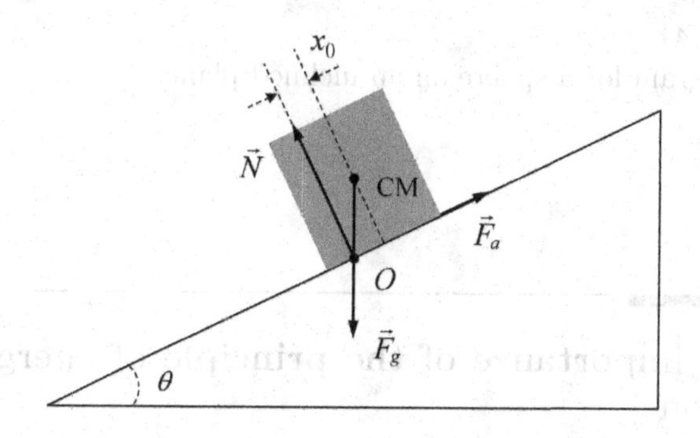

FIGURE 17.3
Showing where the normal force actually acts.

Now the normal force must act *on the body*, so x_0 cannot be greater than $b/2$, which gives a critical value of θ: $\theta_c = \arctan(b/h)$. The body will stay upright if $\theta < \theta_c$, and topple if $\theta > \theta_c$. Students can relax now. We have found the obvious: some furniture will topple, some will not. Besides, Fig. 17.4 provides us with a graphical method for deciding which will happen: for a given item of furniture (h and b) and a given inclination (θ).

The foregoing discussion also makes clear why it is not possible to balance a sphere on a slope, no matter how small θ might be. Indeed, a perfect sphere and a perfect plane have only one point of contact, so there is no freedom when drawing the line of action of the normal force: it must pass through the CM (as in Fig. 17.1) and the sphere will always rotate (see Fig. 17.4).

---

particular x ($= x_0$) for which $\alpha = 0$. So as not to ruin the game, we leave it to the very end to show that the principle of energy conservation actually forces nature to pick this particular x.

[3]This is the obvious solution to the problem. Indeed, for this choice, the lines of action of all forces intersect at O, so the net torque vanishes, and there is no rotation.

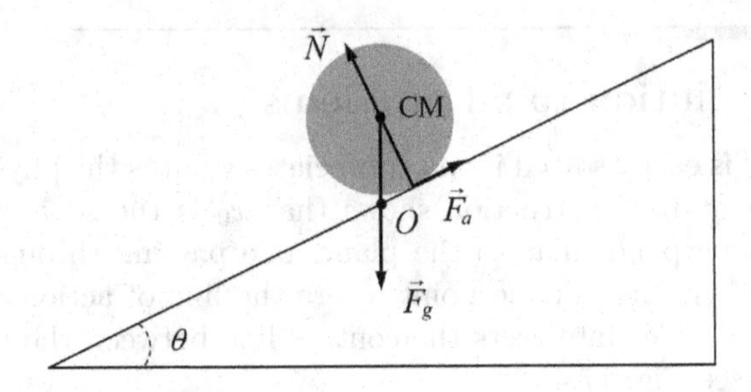

FIGURE 17.4
Free-body diagram for a sphere on an inclined plane.

17.5 The importance of the principle of energy conservation

We have left to the very end a relevant question, so as not to upset the natural course of a teaching session. So far, we have shown that, for $\theta < \theta_c$, we can always find some x ($x = x_0$) for which the body will not topple. However, we have not shown that nature fulfills this condition (although to do otherwise would entail the contradictions discussed above). An elegant way to prove this is to use the principle of energy conservation [4].

Let us consider the schematic in Fig. 17.5 and assume that $\theta < \theta_c$ but nature somehow chooses $x < x_0$. Now the body will rotate about point A. However, because the body is starting from rest, its acceleration is purely tangential. Then the body's kinetic energy will increase because its speed is increasing from zero, but its gravitational potential energy will also increase because its CM is moving up. This violates the principle of energy conservation. Things would be even more dramatic if $x > x_0$, in which case the body would rotate about point B. We are led to the conclusion that for $\theta < \theta_c$ one must draw the normal force passing through $x = x_0$; there is no other option. Finally, note that rotation about A is consistent with the principle of

[4]There is an easier way to prove that the only possible choice is x_0. Start by noting that if the body rotates about A (B), then this will be the only point of contact between body and surface, and therefore the normal force must act there (see Fig. 17.5). Consider Fig. 17.5 with $\theta < \theta_c$ and suppose nature chose $x < x_0$. The body would rotate about A and so the normal force would act on A ($x = b/2$), which contradicts our assumption that $x < x_0$. There will be no contradiction only if $x_0 > b/2$ (for $\theta > \theta_c$), in which case the body will inevitably rotate. Likewise, it can be shown that x cannot be greater than x_0. In summary, nature has only two options: $x = x_0$ (if $x_0 < b/2$) or $x = b/2$ (if $x_0 > b/2$).

energy conservation if $\theta > \theta_c$, for in this case the CM will be moving down from the very start. Thus the kinetic energy increases but the gravitational potential energy decreases, which is consistent with the principle of energy conservation,

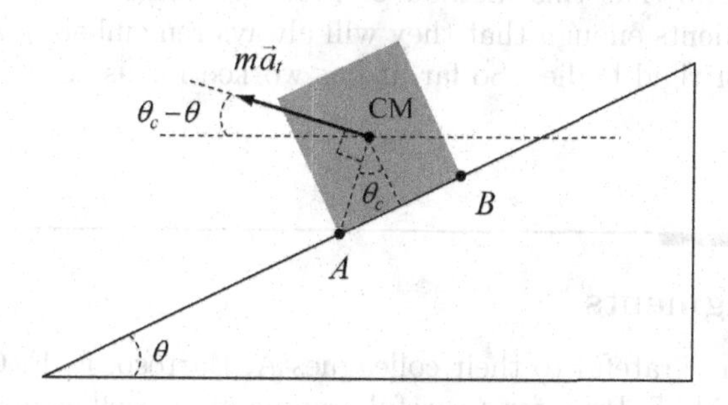

FIGURE 17.5
Here $\theta < \theta_c$. The instantaneous rotation of the body about A at the start of its motion violates the principle of energy conservation.

17.6 Conclusion

We have assumed throughout that the static friction force is strong enough to stop the body sliding down the plane. Working out the correct solution for realistic static and kinetic friction coefficients can be left to a later session [5].

We would like to emphasize that this problem is correctly posed and solved in many mechanics and Year-12 textbooks. However, many students still reach their early (and, except in a few cases, their later) years of higher education in ignorance of where they ought to draw the normal force. In our view the free-body diagram must always be drawn correctly, so we suggest that this "game" should be played as early as Year 12.

On the other hand, it is also clear from our discussion of "the first contradiction," item (a), that the study of translational motion is not affected by this misconception. It is only rotations that "know" about lines of action and are therefore affected. Obviously, when one is only concerned with translational motion, as in Years 10 and 11, it is perfectly acceptable to draw all forces as acting on the body's CM—even friction! When studying rigid body motion, though, Fig. 17.1 is no longer the correct free-body diagram. Approximating a rigid body by a point mass located at its CM where all forces that the body is

[5] See article by Nunes and Silva in this book.

subjected to act, is fine for describing translation but provides no information whatsoever regarding rotation. This misconception arises from confusing two models: one in which the body is approximated as a point mass located at the body's CM, and another in which it is approximated as a rigid body.

It is our hope that this method of proof by contradiction will "shock" and amuse students enough that they will always remember what the correct procedure is for rigid bodies. So far, it has worked for us.

Acknowledgments

The authors are grateful to their colleagues A. Barroso, E. F. Gonçalves, A. M. Nunes and M. T. Peña for a careful reading of, as well as many comments and suggestions on, the manuscript that evolved into this article.

18

Rolling Cylinder on an Inclined Plane

V. Oliveira

CONTENTS

The motion of a cylinder rolling down an inclined plane acted by an external force is studied. The aim is to analyze the influence of the plane's inclination and external force tilting angle on the cylinder's linear and angular accelerations.

18.1 Introduction

The rolling motion of cylinders is an important topic frequently addressed in introductory physics courses for scientists and engineers [1]. It shows up many interesting situations and is very useful to understand concepts such as torque and moment of inertia and the relation between linear and angular accelerations of rigid bodies. As a result, several authors have studied the motion of cylinders acted by horizontal, vertical, or tangential forces [2–5]. However, the purpose of these works was mainly to study the frictional force direction and work, and little attention was given to the direction of the linear and angular accelerations of the cylinder. In a previous paper [6], we presented a systematic study of the motion of a cylinder rolling on a horizontal plane acted by an external force applied at an arbitrary angle, with emphasis on the directions of the linear and angular accelerations of the cylinder. This chapter aims to extend this study by analyzing the motion of the cylinder rolling on an inclined plane.

DOI: 10.1201/9781003187103-18

18.2 Theoretical background

Figure 18.1 shows a cylinder of mass m and radius R rolling in a plane inclined by an angle β. We assume that an external force, F, tilted by an angle θ ($0 \leq \theta \leq 90°$) is applied to the cylinder at a certain height, h. The cylinder rolls acted by this force F, the frictional force, f, generated at the contact point C between the cylinder and the inclined plane, the gravitational force mg and the normal force N. We assume that $F \sin\theta \leq mg\cos\beta$, so movement in the y-direction is not allowed.

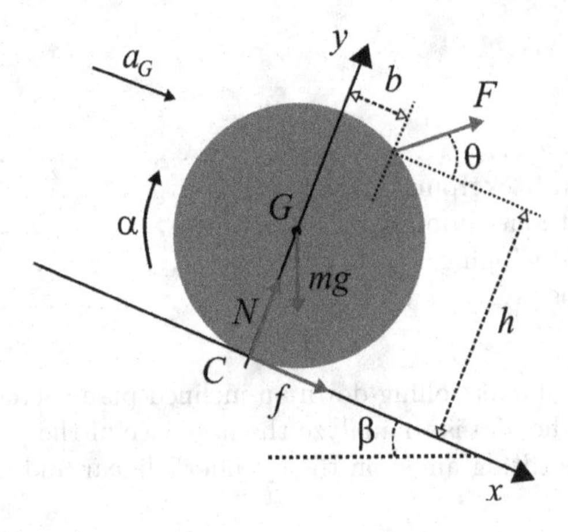

FIGURE 18.1
Schematic representation of a cylinder on an inclined plane.

Applying Newton's second law gives

$$F\cos\theta + f + mg\sin\beta = ma_G, \tag{18.1}$$

and

$$F\sin\theta + N - mg\cos\beta = 0, \tag{18.2}$$

where a_G is the linear acceleration of the center-of-mass, considered positive if pointing in the positive direction of the x-axis. The rotational analog of Newton's second law is

$$F\cos\theta \cdot (h - R) - F\sin\theta \cdot b - f \cdot R = \frac{1}{2}mR^2\alpha, \tag{18.3}$$

where α is the angular acceleration of the cylinder considered positive if clockwise and negative if counter-clockwise, $\frac{1}{2}mR^2$ is the moment of inertia of the cylinder relative to its axis, and $b = h\sqrt{2R/h - 1}$. By combining Eqs. (18.1) and (18.3), we obtain a relation between the linear and angular accelerations:

$$2a_G + \alpha R = \frac{2F}{mR}(h\cos\theta - b\sin\theta) + 2g\sin\beta. \tag{18.4}$$

18.3 Rolling without slipping

If the contact point C is stationary, the cylinder will roll without slipping. As a result, a_G and α are simultaneously positive or negative and related by the condition $a_G = \alpha R$. The linear acceleration is then given by:

$$a_G = \frac{2F}{3mR}(h\cos\theta - b\sin\theta) + \frac{2}{3}g\sin\beta. \tag{18.5}$$

For simplicity, let us write the force F in the form $F = kmg$ where $k \leq \frac{\cos\beta}{\sin\theta}$ is a constant. With this substitution, Eq. (18.5) turns into

$$a_G = \frac{2kg}{3R}(h\cos\theta - b\sin\theta) + \frac{2}{3}g\sin\beta. \tag{18.6}$$

For $k = 0$ ($F = 0$) we retrieve the linear acceleration of a cylinder rolling down an inclined plane, $a_G = \frac{2}{3}g\sin\beta$. In this case, both the linear and angular accelerations are positive. For $k > 0$ ($F > 0$), the direction of these accelerations will depend on θ and β. Figure 18.2 shows the linear acceleration of the cylinder as a function of the angles θ and β for $k = 0.3$ and $h/R = 0.5, 1$, and 1.5. For $\beta = 0$ we obtain the linear acceleration for an horizontal plane [6]. In this case, there is a critical angle $\theta_c = \tan^{-1}\left(\frac{h}{b}\right)$ for which a_G and α are null. For the selected h/R ratios, the critical angles are 30°, 45°, and 60°, respectively. For $\theta = \theta_c$, the cylinder cannot roll without slipping. On the other hand, the linear acceleration is positive for $\theta < \theta_c$, and negative for $\theta > \theta_c$. For $\beta > 0$, the existence of a critical angle depends on the ratio h/R and the value of β. For example, for $h/R = 1$ and $\beta = 7.5°$, the critical angle is about 62.9°. The linear acceleration is negative above this critical angle and positive below it. On the other hand, for $\beta = 30°$ and 60°, the linear acceleration is always positive. A similar behavior is observed for $h/R = 1$ and $h/R = 1.5$. Note that, for negative accelerations, the cylinder goes up the inclined plane.

Since the frictional force f is static, it has a maximum value given by $|f| \leq \mu_s N$, where μ_s is the coefficient of static friction. From Eq. (18.2) we can write $|f| \leq \mu_s mg(\cos\beta - k\sin\theta)$, and, taking into account Eqs. (18.1) and (18.6), it follows that rolling without slipping occurs when

$$\mu_s \geq \frac{\left|\frac{2k}{3R}[(h - \frac{3}{2}R)\cos\theta - b\sin\theta] - \frac{1}{3}\sin\beta\right|}{\cos\beta - k\sin\theta}. \tag{18.7}$$

The minimum value of μ_s that guarantees rolling without slipping is plotted in Fig. 18.3 for $k = 0.3$. For $h/R = 1.5$, it increases with β and θ. On the other hand, for $h/R = 0.5$ and 1.0, it increases with β, but reaches a maximum for a certain θ. It is worth analysing the case of $h = 3R/2$, $\beta = 0$ (horizontal plane) and $\theta = 0$ (horizontal force). For this particular situation, we have $\mu_s \geq 0$, and the cylinder will always roll without slipping, irrespective of the friction coefficient value.

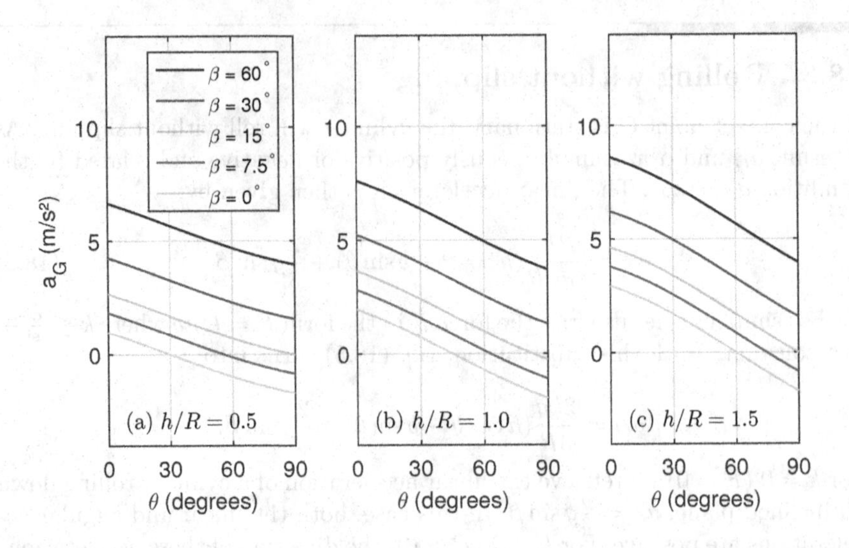

FIGURE 18.2
Linear acceleration of the cylinder as a function of θ and β for $k = 0.3$ and (a) $h/R = 0.5$; (b) $h/R = 1$; and (c) $h/R = 1.5$.

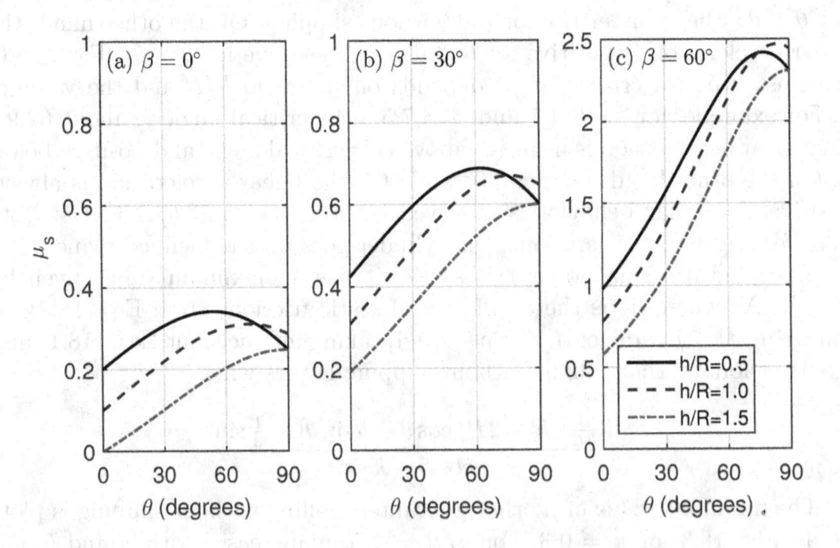

FIGURE 18.3
Minimum static friction coefficient for rolling without slipping as a function of θ and h/R for $k = 0.3$ and (a) $\beta = 0°$; (b) $\beta = 30°$; and (c) $\beta = 60°$.

18.4 Rolling and slipping

If the velocity of the contact point C is not zero, the frictional force is kinematic and has a constant value given by

$$|f| = \mu_k mg(\cos\beta - k\sin\theta), \tag{18.8}$$

where μ_k is the coefficient of kinematic friction, and the use of Eq. (18.2) has been made. From Eqs. (18.1) and (18.3), we get the following equation for the linear and angular accelerations,

$$a_G = g(\sin\beta + \mu_k\cos\beta) + gk(\cos\theta - \mu_k\sin\theta), \tag{18.9}$$

and

$$\alpha R = 2kg[(h/R - 1)\cos\theta + (\mu_k - b/R)\sin\theta] - 2\mu_k g\cos\beta. \tag{18.10}$$

Figures 18.4 to 18.6 show plots of a_G and αR as a function of θ, for $\mu_k = 0.1$ and different values of β, h/R and k. The linear acceleration is always positive and increases with β (Fig. 18.4) and k (Fig. 18.6), but it does not depend on the ratio h/R (Fig. 18.5). On the other hand, the angular acceleration does not depend significantly on β (Fig. 18.4), but it does depend on h/R (Fig. 18.5) and k (Fig. 18.6). In contrast to rolling without slipping, the plots also show

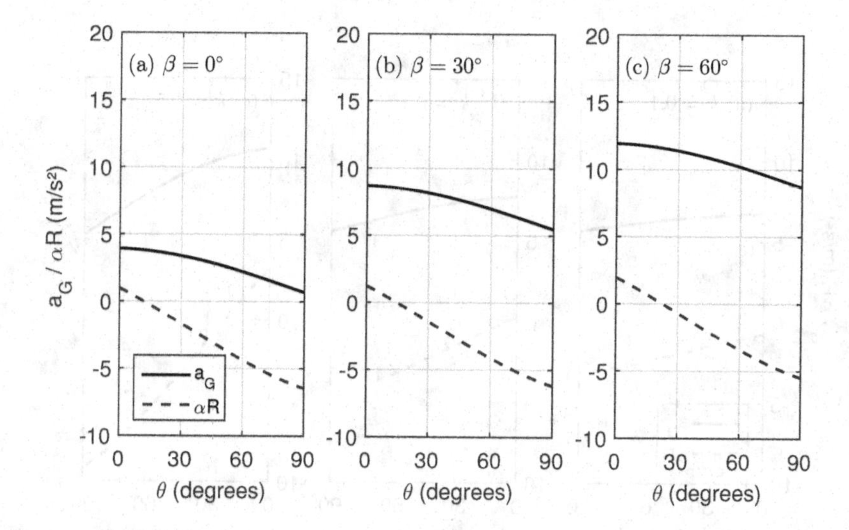

FIGURE 18.4
Plots of a_G (continuous line) and αR (dashed line) as a function of θ, for $k = 0.3$, $h/R = 1.5$, $\mu_k = 0.1$ and (a) $\beta = 0°$; (b) $\beta = 30°$; and (c) $\beta = 60°$.

that the linear and angular acceleration may have opposite signs. The linear acceleration is always positive (downward the inclined plane), but the angular acceleration may be positive (clockwise), negative (counter-clockwise), or null depending on the values of θ, k, h/R and β. It is interesting to note that for $\alpha = 0$, the cylinder goes down the inclined plane without rolling.

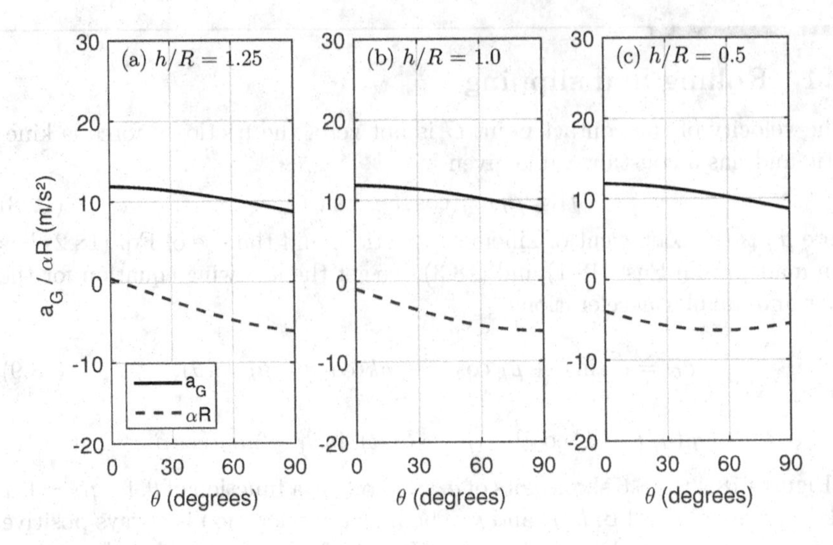

FIGURE 18.5
Plots of a_G (continuous line) and αR (dashed line) as a function of θ, for $k = 0.3$, $\beta = 30°$, $\mu_k = 0.1$ and (a) $h/R = 1.25$; (b) $h/R = 1.0$; and (c) $h/R = 0.5$.

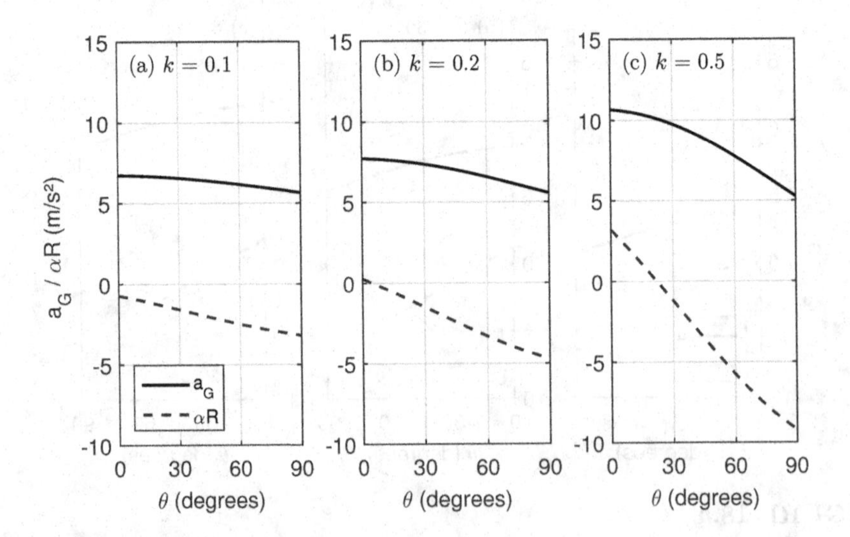

FIGURE 18.6
Plots of a_G (continuous line) and αR (dashed line) as a function of θ, for $\beta = 30°$, $h/R = 1.5$, $\mu_k = 0.1$ and (a) $k = 0.1$; (b) $k = 0.2$; and (c) $k = 0.5$.

It is also particularly interesting to plot Eqs. (18.9) and (18.10) for a frictionless surface ($\mu_k = 0$). This is shown in Fig. 18.7 for $\beta = 0$, $30°$, and $60°$. In this case, the angle β does not affect the angular acceleration of the cylinder. On the other hand, a_G is null for $\beta = 0°$ and $\theta = 90°$. For this case, the cylinder will slip in place counter-clockwise ($\alpha < 0$) and does not roll. Finally, for $\beta = \theta = 0°$, the cylinder does not slip since we have $a_G = \alpha R$.

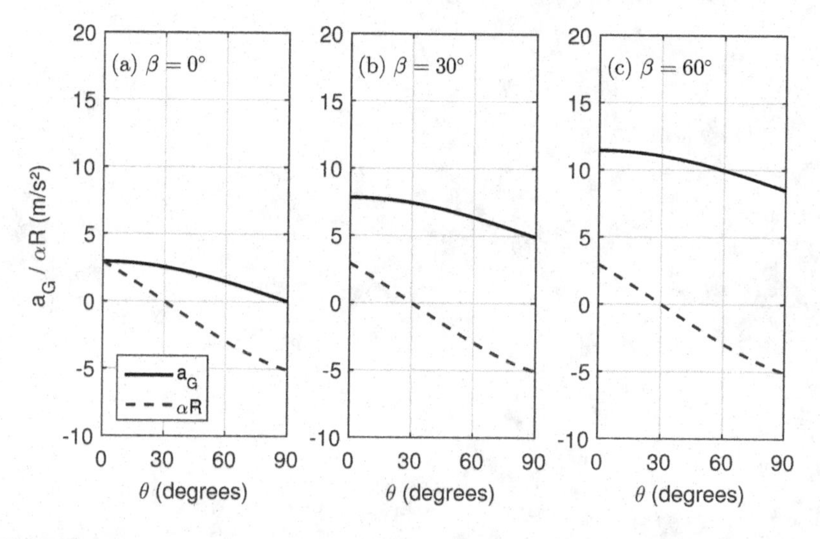

FIGURE 18.7
Plots of a_G (continuous line) and αR (dashed line) as a function of θ, for a frictionless surface and $k = 0.3$, $h/R = 1.5$ and (a) $\beta = 0°$; (b) $\beta = 30°$; and (c) $\beta = 60°$.

18.5 Conclusions

The motion of a cylinder rolling down an inclined plane acted by an external force is studied. If rolling occurs without slipping, the linear and angular accelerations can be positive or negative but cannot have opposite signs. As a result, in this case, the cylinder may go up or down the inclined plane. In contrast, if rolling occurs with slipping, the accelerations may have opposite signs. For example, the linear acceleration may be positive and the angular acceleration negative. For this situation, the cylinder goes down the inclined plane rolling counter-clockwise.

References

[1] F. P. Beer, E. R. Johnston, and W, E. Clausen, *Vector Mechanics for Engineers: Dynamics*, 7th ed. (McGraw-Hill, New York, NY, 2004).

[2] A. Salazar, A. Sánchez-Lavega, and M. A. Arriandiaga, "Is the frictional force always opposed to the motion?," Phys. Educ. **25**, 82–85 (1990).

[3] C. Carnero, J. Aguiar, and J. Hierrezuelo, "The work of the frictional force in rolling motion," Phys. Educ. **28**, 225–227 (1993).

[4] A. Pinto and M. Fiolhais, "Rolling cylinder on a horizontal plane, " Phys. Educ. **36**, 250–260 (2001).

[5] C. E. Mungan, "Acceleration of a pulled spool," Phys. Teach. **39**, 481–485 (2001).

[6] V. Oliveira, "Angular and linear accelerations of a rolling cylinder acted by an external force," Eur. J. Phys. **32**, 381–388 (2011).

Index

Printed in the United States
by Baker & Taylor Publisher Services

Printed in the United States
by Baker & Taylor Publisher Services